# Probability and Statistics Explorations with Maple

Strong
H 529-5705
ECON

# Probability and Statistics Explorations with Maple


Zaven A. Karian

*Denison University*
*Granville, Ohio*

Elliot A. Tanis

*Hope College*
*Holland, Michigan*

*Prentice-Hall, Inc., Englewood Cliffs, New Jersey 07632*

Production Editor: *Joan Eurell*
Acquisitions Editor: *Ann Heath*
Supplement Acquisitions Editor: *Audra Walsh*
Production Coordinator: *Alan Fischer*

A Paramount Communications Company
Englewood Cliffs, New Jersey 07632



Printed in the United States of America

10 9 8 7 6 5 4 3 2 1

ISBN 0-13-436759-6

Prentice-Hall International (UK) Limited, *London*
Prentice-Hall of Australia Pty. Limited, *Sydney*
Prentice-Hall Canada Inc., *Toronto*
Prentice-Hall Hispanoamericana, S.A., *Mexico*
Prentice-Hall of India Private Limited, *New Delhi*
Prentice-Hall of Japan, Inc., *Tokyo*
Simon & Schuster Asia Pte. Ltd., *Singapore*
Editora Prentice-Hall do Brasil, Ltda., *Rio de Janeiro*

# Contents

# Preface

The development of statistical packages such as SPSS and BMPD has altered the practice of statistics during the past two decades. Such systems tend to have powerful numeric and graphic features but lack symbolic computing capabilities. By contrast, computer algebra systems provide a rich environment for mathematical manipulations but lack many features that are essential for statistical analysis and for probability and statistics instruction. The supplement of approximately 130 procedures to the *Maple* computer algebra system that we have developed gives us both worlds: some statistical analysis and also explorations within a rich mathematical environment.

*Maple*, this laboratory manual, and the supplementary *Maple* procedures provide the basic tools for enhancing student learning in a calculus-based probability and statistics course. In biology, synergism occurs when two (or more) substances or organisms achieve an effect of which each is individually incapable. We believe that *Maple*, with our statistical enhancements, has similar potential for student understanding and learning. The central objective of these laboratory exercises and the accompanying *Maple* procedures is to improve students' understanding of basic concepts. This is done in several ways.

**Simulations**: Some concepts in probability and statistics can be illustrated with random experiments that use such things as dice, cards, coins, and urns containing colored balls. A few of the exercises make use of such devices. However the nature of such experiments limits their applicability and often dictates that sample sizes be small. The exercises emphasize the use of computer simulations. *Maple* procedures are provided for both graphical and numerical comparisons between empirical (simulated) data and theoretical models.

**Explorations of Models**: When developing probability models, it is important to see and understand the shape of each probability distribution. It is also necessary to be able to calculate characteristics of each distribution such as its mean, variance, and distribution function. *Maple*, together with the statistical supplement that accompanies this book, provides the tools for doing this.

**Development of Insight**: It is sometimes possible to go through the steps of a proof of a theorem without really understanding the implications of the theorem or the importance of all of the assumptions that are used for the proof. Empirically it is possible to see, for example, how the sample size, $n$, affects the conclusion of the Central Limit Theorem. It is possible to get a feeling for the distribution of $T = (\overline{X} - \mu)/(\sigma/\sqrt{n}\,)$ when the distribution from which the sample is taken is **not** normal.

**Statistical Computations**: Given a set of measurements, it is possible to explore characteristics of the data without being overwhelmed with tedious calculations. Assumptions about the data can be checked. For example, it is possible to test whether the data come from a normal distribution or perhaps from a skewed distribution.

We have used these materials in scheduled laboratory sessions that meet concurrently with a course as well as in circumstances where the instructor uses these problems as "take-home" laboratory exercises. Selected exercises can be used to supplement any mathematical statistics or probability course as take-home assignments, laboratory exercises, or classroom demonstrations. Selected exercises can also be used for student independent study projects.

The important thing is to use these materials as a starting point. Hopefully some exercises and their solutions will suggest further questions and open new areas of exploration. Sample sizes are given as guides. Depending on your particular computer, you may want to increase or decrease the suggested numbers. Experiment and share your successes and failures with the authors.

We have tried to minimize the extent to which students and faculty need to develop expertise in *Maple* in order to take advantage of the exercises. Appendix A is a very brief and general introduction to *Maple*. Although incomplete, it provides a starting point for novices to begin using *Maple*. Appendix B gives a reasonably detailed description of the procedures of our statistics supplement. We should note here that a standard *Maple*-like help system is bundled with the supplement so that a user can obtain on-line help at any time during a session. The supplement and associated files (help files, data files, installation instructions) can be obtained either over the internet through the ftp utility or on a diskette from the publisher. The latter, which can be in either DOS or Macintosh format, includes solutions to all of the exercises in this book. However you obtain these materials, start by reading the README file.

The sequence of commands given below can be used to obtain the *Maple* supplement via the internet. Following the first command, use "anonymous" for your username and your e-mail address for your password.

```
ftp   ftp.maplesoft.on.ca
cd   pub/maple/books/karian
mget   *.*
binary
get   stat.m
bye
```

This text and associated materials can be used with almost all mathematical statistics books. However, the order of the topics parallels those in *Probability and Statistical Inference*, 4th edition, by Robert V. Hogg and Elliot A. Tanis (Macmillan Publishing Company, 1993—currently available from Prentice-Hall, Inc.). Most of the data in the exercises of this book are also included with the distribution disk from the publisher and the ftp files.

Many people have been involved in the preparation of these materials. Those who have had the most influence are our students who have helped with writing computer software and have reacted to exercises as they have been written. We thank each of them, even though it is not possible to mention all of them by name. Two Hope College students and one Denison University student have played key roles during the latest revision in which *Maple* has been incorporated. We thank Bryan R. Goodman and Joshua D. Levy of Hope College and Rohit Goyal of Denison University for their outstanding work. We wish to acknowledge the Curtis A. Jacobs Memorial Fund of Hope College and the Faculty Development Program of Denison University for providing support for this project. Stan Devitt, Development Manager of Waterloo Maple Software, helped us with a number of technical problems and facilitated the distribution of the *Maple* supplement via ftp; we are indebted to him. We also thank Lori McDowell for typing several drafts of this text in LaTeX.

The authors would appreciate receiving comments and suggestions.

Zaven A. Karian
Denison University
Granville, Ohio
E-Mail: Karian@Denison.edu

Elliot A. Tanis
Hope College
Holland, MI
E-Mail: Tanis@Math.Hope.edu

# Chapter 1

# Summary and Display of Data

## 1.1 Random Number Generators

Many of the exercises in this computer-based laboratory manual depend on a random number generator. Theoretically, a random number generator selects a real number at random from the interval [0,1). That is, each number in the interval [0,1) has the same chance of being selected. For various reasons, each of the methods described below produces pseudo-random numbers. However, the numbers that are generated usually behave like numbers selected at random from the interval [0,1). For this reason, and for convenience, we shall drop the adjective "pseudo" and the generated numbers will be called random numbers.

In addition to numbers selected randomly from the interval [0,1), there is often interest in selecting a set of integers randomly from some interval. Methods for doing this are also described.

**Method 1:** Tables of random numbers that have been generated by various methods are available in many statistics books. To use a table of random numbers such as the one given in Appendix C, select a starting value at random and then proceed systematically to obtain the desired number of random numbers.

**Method 2:** The *Maple* function `rng();` can be used to simulate the random selection of a number from the interval $[0, 1)$. The function `RNG(n)` in the *Maple* supplement simulates the random selection of `n` numbers from the interval $[0, 1)$ and returns them as a list. For example, `X := RNG(10);` will set `X` equal to a list of 10 random numbers.

The *Maple* function `rand();` can be used to produce random integers between 0 and 999999999988 (almost $10^{12}$). It can also be used to produce random integers on a specified integer range. For example, `rint := rand(10 .. 20); rint();` will select an integer randomly from the interval 10 to 20, inclusive. Additionally, the *Maple* supplement has a function `Die(m,n)` that will simulate a list of `n` integers that are selected randomly between 1 and `m`, inclusive. For example, `Y := Die(6,20);` will simulate 20 rolls of a 6-sided die and return them in a list named `Y`.

**Remark.** If you want to obtain different results on successive interactions with *Maple*, issue the `randomize():` command at the beginning of your session.

**Method 3:** Most calculators have a built in random number generator that can be used to simulate the random selection of integers or numbers from the interval [0,1).

## EXERCISES

**Purpose:** The exercises illustrate the three methods for generating random numbers, helping you to become familiar with outcomes of random experiments. Also some simple random experiments are simulated.

**1.1–1** Select a set of 30 random numbers from the table in Appendix C.

(a) How many of your observations fell in each of the intervals $A_1 = [0.00, 0.20)$, $A_2 = [0.20, 0.40)$, $A_3 = [0.40, 0.60)$, $A_4 = [0.60, 0.80)$, $A_5 = [0.80, 1.00)$?

(b) What proportion of the 30 observations would you have expected to fall in each of the intervals?

(c) Answer parts (a) and (b) if the intervals are $C_1 = [0.00, 0.50)$, $C_2 = [0.50, 1.00)$?

(d) Answer parts (a) and (b) if the intervals are $D_1 = [0, 1/6)$, $D_2 = [1/6, 1.00)$.

**1.1–2** Use `RNG` to generate 30 random numbers and answer the four questions in Exercise 1.1–1 for these 30 numbers. The numbers can be produced by `R := RNG(30);` which, if followed by `SR := sort(R);`, will arrange the random numbers in non-decreasing order.

**1.1–3** Use a calculator random number generator to simulate the selection of 30 random numbers. Answer the four questions in Exercise 1.1–1 for these 30 numbers.

**1.1–4** A tetrahedron (4-sided die) has 4 possible outcomes, 1, 2, 3, 4, that occur with equal probability. Use `T := Die(4,n);` to simulate $n$ rolls of a tetrahedron.

(a) Simulate 400 rolls of a fair tetrahedron.

(b) Use the function `Freq` to print the number of times each outcome was observed. Since the values of `T` have to be between 1 and 4, `Freq(T,1 .. 4);` will give the successive frequencies of 1, 2, 3, and 4. Compare and combine your results with those of other members in the class. Are the results consistent with what you expected?

**1.1–5** An octahedron (8-sided die) has 8 possible outcomes, 1, 2, 3, 4, 5, 6, 7, 8, that occur with equal probability.

(a) Simulate a single roll of an 8-sided die.

(b) Simulate 400 rolls of a fair 8-sided die.

(c) Print the number of times each outcome was observed (use `Freq` to do this). Compare and combine your results with those of other members in the class. Are the results consistent with what you expected?

**1.1–6** Simulate the experiment of flipping a coin until the same face is observed on successive flips. What is the average number of flips needed to observe the same face? (See Exercise 3.6-18.)

Coin tosses, say 20 of them, can be simulated by `Flips := Die(2,20);`. If `A` is defined by `A := [seq(Flips[i]-Flips[i-1],i = 2 .. 20)];`, then a zero entry in `A` indicates a repetition and `B := Locate(A,0);` gives the locations in `A` where 0's occur. To obtain the number of flips until the same face is observed on successive flips, we use `C := [seq(B[i] - B[i-1] + 1,i = 2 .. nops(B))];` (`nops(B)` gives the number of elements in `B`). Now `Mean(C);` will give the average.

It is also possible to use the programming features of *Maple*, via the construction of loops, to obtain the same result. This is shown below for the simulation of 500 coins.

```
randomize();
for i to 500 do
old_coin := [0];
new_coin := Die(2,1);
XX[i] := 1;
while new_coin <> old_coin do
old_coin := new_coin;
new_coin := Die(2,1);
XX[i] := XX[i]+1
od
od;
X := [seq(XX[i],i = 1 .. 500)];
m := Mean(X);
```

**1.1–7** Simulate the experiment of flipping a coin until heads is observed on successive flips. What is the average number of flips needed to observe successive heads? (See Exercise 3.6-19.) **Hint:** Replace `while new_coin <> old_coin do` with `while op(new_coin) + op(old_coin) < 4 do` in the solution for the last exercise.

**Remark:** Note that `Die(2,1)` can equal `[1]` or `[2]` while `op(Die(2,1))` can equal 1 or 2. Thus, if 2 represents a head, then a sum of 4 would indicate the observance of successive heads.

**1.1–8** * This exercise yields an interesting display.

(a) Plot three points, e.g., $P = (1, 1)$, $Q = (50, 95)$, and $R = (99, 1)$.

(b) Select a point randomly with $x$ between 0 and 100 and $y$ between 0 and 100. Call this point (`A[1], B[1]`).

(c) Randomly move 1/2 the distance from the current point (`A[1], B[1]`) to $P, Q$, or $R$. That is, select one of the integers 1, 2, and 3 randomly (each with probability 1/3). If 1 is selected, move 1/2 the distance to $P$, if 2 is selected, move 1/2 the distance to $Q$, and if 3 is selected, move 1/2 the distance to $R$. Store the new point.

(d) Repeat part (c) 500 or more times with (`A[i], B[i]`) the new point.

(e) Plot these points using `ScatPlot;`.

```
A[1] := 100*rng();
B[1] := 100*rng();
m := 500;
for i from 2 to m do
roll := Die(3,1);
if roll = [1] then
A[i] := 1/2*A[i-1]+1/2;
B[i] := 1/2*B[i-1]+1/2
elif roll = [2] then
A[i] := 1/2*A[i-1]+25;
B[i] := 1/2*B[i-1]+95/2
else
A[i] := 1/2*A[i-1]+99/2;
B[i] := 1/2*B[i-1]+1/2
fi
od;
X := [seq(A[i],i = 1 .. m)];
Y := [seq(B[i],i = 1 .. m)];
ScatPlot(X,Y);
```

**1.1–9** Repeat Exercise 1.1–8 with different points $P$, $Q$, and $R$.

## Questions and Comments

**1.1–1** Compare the results for Exercises 1.1–1, 1.1–2, and 1.1–3. Do the three random number generators behave similarly?

**1.1–2** If 500 random numbers had been generated in Exercise 1.1–2, would you have expected the numbers of outcomes and/or the proportions of outcomes in $A_1$, $A_2$, $A_3$, $A_4$, and $A_5$ to be closer to each other? Why?

## 1.2 Samples, Histograms, and Ogives

A random experiment associated with a random variable $X$ is repeated $n$ independent times. Let $X_1, X_2, \ldots, X_n$ denote the $n$ random variables associated with the $n$ trials. These random variables are called a random sample of size $n$. They are independent and each has the same distribution as $X$. The observed values of the $n$ random variables in the random sample are denoted by $x_1, x_2, \ldots, x_n$.

The sample mean, or mean of the empirical distribution, is defined by

$$\overline{X} = \frac{1}{n}\sum_{i=1}^{n} X_i.$$

The observed value of the sample mean is

$$\overline{x} = \frac{1}{n}\sum_{i=1}^{n} x_i.$$

The variance of the empirical distribution, is defined by

$$V = \frac{1}{n}\sum_{i=1}^{n}(X_i - \overline{X})^2.$$

The observed value of the variance of the empirical distribution is

$$v = \frac{1}{n}\sum_{i=1}^{n}(x_i - \overline{x})^2 = \frac{1}{n}\sum_{i=1}^{n} x_i^2 - \overline{x}^2.$$

The sample variance is

$$S^2 = \frac{1}{n-1}\sum_{i=1}^{n}(X_i - \overline{X})^2 = \frac{nV}{n-1}.$$

The observed value of the sample variance is

$$s^2 = \left(\frac{1}{n-1}\right)\sum_{i=1}^{n}(x_i - \overline{x})^2 = \frac{n\sum_{i=1}^{n} x_i^2 - (\sum_{i=1}^{n} x_i)^2}{n(n-1)}.$$

Note that

$$s^2 = [n/(n-1)]v \quad \text{and} \quad v = [(n-1)/n]s^2.$$

The respective standard deviations are $\sqrt{v}$ and $s = \sqrt{s^2}$.
Chebyshev's inequality for empirical distributions states that, for every $k \geq 1$,

$$\frac{\#(\{x_i : |\, x_i - \overline{x} \,| \geq k\sqrt{v}\})}{n} \leq 1/k^2$$

or equivalently,

$$\frac{\#(\{x_i : |\, x_i - \overline{x}\,| < k\sqrt{v}\})}{n} \geq 1 - 1/k^2.$$

These inequalities hold if $\sqrt{v}$ is replaced by $s$.

The empirical distribution function is defined by

$$F_n(x) = \#(\{x_i :\ x_i \leq x\})/n.$$

Note that in the empirical distribution, a weight (empirical probability) of $1/n$ is assigned to each observed $x_i$, $i = 1, 2, \ldots, n$. Thus, we have a distribution of the discrete type.

Suppose that a random experiment is such that the space of the random variable $X$ associated with this experiment is $R = \{x :\ a \leq x \leq b\}$. If this experiment is repeated $n$ independent times with observations $x_1, x_2, \ldots, x_n$, the relative frequency $\#(\{x_i :\ c < x_i \leq d\})/n$ can be used as an estimate of $P(c < X \leq d)$, $a \leq c < d \leq b$. The relative frequency histogram extends this idea.

Let $a = c_0 < c_1 < c_2 < \cdots < c_k = b$. Let $f_j = \#(\{x_i :\ c_{j-1} < x_i \leq c_j\})$ denote the frequency or number of outcomes in the interval $(c_{j-1}, c_j]$, $j = 1, 2, \ldots, k$. The function defined by

$$h(x) = \frac{f_j}{n(c_j - c_{j-1})}, \quad c_{j-1} < x \leq c_j, \quad j = 1, 2, \ldots, k,$$

is called a relative frequency histogram. Note that

$$\int_{c_{j-1}}^{c_j} h(x)dx = (f_j/n) = \frac{\#(\{x_k :\ c_{j-1} < x_i \leq c_j\})}{n}$$

is the relative frequency of outcomes in the interval $(c_{j-1}, c_j]$. Furthermore,

$$\int_a^b h(x)dx = 1.$$

Let $F_n(x)$ denote the empirical distribution function. In the above notation

$$F_n(c_j) = \frac{\#(\{x_i :\ x_i \leq c_j\})}{n} = \frac{\sum_{i=1}^{j} f_i}{n}.$$

The graph obtained by plotting the points $[c_0, F_n(c_0)]$, $[c_1, F_n(c_1)]$, $\ldots$, $[c_k, F_n(c_k)]$ and drawing line segments between each pair of adjacent points is called a relative frequency ogive curve. We shall denote this function by $H(x)$.

The sample mean, $\overline{x}$, will generally become closer to the distribution or population mean, $\mu$, as the sample size increases. That is, the running averages, $x_1$, $(x_1+x_2)/2$, $(x_1+x_2+x_3)/3$, $\ldots$, $(x_1 + x_2 + \cdots + x_n)/n$, can be viewed as successively more refined estimates of the true mean $\mu$.

## EXERCISES

**Purpose:** The exercises illustrate the relation between sample and distribution characteristics. Applications of `Histogram`, `Ogive`, `PlotEmpCDF`, and `PlotRunningAverage` for depicting the histogram, ogive, empirical distribution function, and a graph of successive sample means are illustrated. The use of the random number generator for simulating a problem is considered.

**1.2–1** Use `X := RNG(500);` to generate 500 random numbers between 0 and 1.

(a) Find the values of $\overline{x}$ , $v$ and $s^2$. Is $\overline{x}$ approximately equal to $\mu = 1/2$? Is $s^2$ approximately equal to $\sigma^2 = 1/12$?

```
randomize();
X := RNG(500);
xbar := Mean(X);
s2 := Variance(X);
v := 499/500*s2;
```

(b) Use `PlotRunningAverage(X);` to obtain a graph of the successive running averages. Do the running averages converge to $\mu = 0.5$?

```
p1 := PlotRunningAverage(X):
y2 := 0.5:
p2 := plot(y2,x = 0 .. 500):
plot({p1,p2});
```

(c) Use `Histogram` to depict a relative frequency histogram along with the theoretical probability density function (p.d.f.) $f(x) = 1,\ 0 \le x \le 1$, superimposed.

```
p3 := Histogram(X,0 .. 1,20):
y4 := 1;
p4 := plot(y4,x = 0 .. 1):
plot({p3,p4});
```

(d) Use `Ogive` to depict the ogive curve along with the distribution function $F(x) = x,\ 0 \le x \le 1$, superimposed.

(e) Plot the empirical distribution function with the distribution function.

```
p5 := Ogive(X,0 .. 1,20):
y6 := x:
p6 := plot(y6,x = 0 .. 1):
plot({p5,p6});
p7 := PlotEmpCDF(X, 0 .. 1):
plot({p6,p7});
```

(f) What proportion of your observations fell within $(\overline{x} - 1.5\sqrt{v},\ \overline{x} + 1.5\sqrt{v})$? How does this proportion compare with the bound guaranteed by Chebyshev's Inequality? You may use

```
num := 0;
for k from 1 to 500 do
if abs(X[k]-xbar) < 1.5*sqrt(v) then
num := num+1
fi
od;
num/500;
```

or

```
U := [seq(abs(X[i]-xbar),i = 1 .. 500)];
US := sort(U);
c := 1.5*sqrt(v);
US[425 .. 445];
```

and then make some observations.

**1.2–2** Let the random variable $X$ be defined by `X := 2*sqrt(rng());`, where `rng()` represents a random number. Generate 200 observations of $X$. This can be done by

```
Y := RNG(200);
X := [seq(2*sqrt(Y[i]),i = 1 .. 200)];
```

(a) Find the values of $\overline{x}$ and $s^2$. Is $\overline{x}$ approximately equal to $\mu = 4/3$? Is $s^2$ approximately equal to $\sigma^2 = 2/9$?

(b) How well do the running averages converge to $\mu = 4/3$?

(c) Use `Histogram` to depict a relative frequency histogram along with the probability density function (p.d.f.) $f(x) = x/2,\ 0 \le x \le 2$, superimposed.

(d) Use `Ogive` to depict the ogive curve along with the distribution function $F(x) = x^2/4, 0 \le x \le 2$, superimposed.

(e) Use `PlotEmpCDF` to plot the empirical distribution function with the theoretical distribution function superimposed.

(f) How good is the "fit" in parts (c), (d), and (e)?

**1.2–3** It is advantageous to graph data like that simulated in Exercise 1.1-5 to "see the shape" of the data. Simulate 400 observations of the roll of an 8-sided die. You may use `D := Die(8,400):`. Then use `Histogram(D,0.5 .. 8.5,8);` to graph these data. Find the frequencies of each of the outcomes using `Freq(D,1 .. 8);`.

**1.2–4** Use `Histogram`, `Ogive`, `PlotEmpCDF`, and `PlotRunningAverage` to graph sets of data from your textbook or from some other source. Also use `Mean`, `Variance`, and `StDev` to find the means, variances, and standard deviations of your data.

**1.2–5** A salesperson makes periodic visits to each of a number of customers who bottle soft drinks. The salesperson sells them glass bottles by the case. The number of cases ordered by one of these regular customers is approximately uniformly distributed from 0 to 999 cases. That is, the probability is approximately 1/1000 that this customer orders $k$ cases, $k = 0, 1, \ldots, 999$. The salesperson makes a commission of 10 cents on each of the first 400 cases in an order and 12 cents on each case above the first 400.

(a) Describe how you would simulate the number of cases sold to a customer. Now describe how you would simulate the random variable, $X$, that represents the commission in dollars on an order from a regular customer.

(b) Simulate 50 random observations of $X$ and use `PlotEmpCDF` to depict the empirical distribution function.

**1.2–6** The salesperson in Exercise 1.2–5 discovers that it is possible to visit 1, 2, or 3 regular customers on any particular day with probabilities 0.3, 0.5, and 0.2, respectively. The numbers of cases ordered by the customers are mutually independent and each has the given uniform distribution.

(a) Describe how the random number generator can be used to simulate the random variable that gives the salesperson's commission per day.

(b) Simulate a random sample of size 30 from this distribution. Use `PlotEmpCDF` to depict the empirical distribution function.

(c) Calculate the average commission based on your 30 observations. Is this average close to what you would expect the salesperson to receive on the average per day?

## Questions and Comments

**1.2–1** Show that the derivative of the ogive curve, $H'(x)$, is equal to $h(x)$, the relative frequency histogram, whenever the derivative exists.

**1.2–2** Describe the relation between the cumulative area under the relative frequency histogram and the relative frequency ogive curve.

**1.2–3** What is the value of the total area of the rectangles in a relative frequency histogram?

**1.2–4** Which seems to be closer, $\overline{x}$ to $\mu$ or $s^2$ to $\sigma^2$?

**1.2–5** Treating $\overline{X}$ and $S^2$ as random variables, which seems to be larger, $\mathrm{Var}(\overline{X})$ or $\mathrm{Var}(S^2)$?

## 1.3 Exploratory Data Analysis

Given a set of data, one way to present it graphically is with a stem-and-leaf display. Such a display has the same effect as a histogram but the original numbers are not lost.

The first and $n$th order statistics along with the first and third quartiles and the median can be used to construct a box-and-whisker display. Such a display gives a good indication of the skewness and possible outliers of a set of data.

### EXERCISES

**Purpose:** The exercises illustrate stem-and-leaf and box-and-whisker displays using both generated data and observed data sets.

**1.3–1** Simulate 100 observations from each of the following distributions. For each sample construct a stem-and-leaf display and a box-and-whisker display using the subroutines `StemLeaf` and `BoxWhisker`. Compare the displays for these distributions and interpret your output.

- Observations from the exponential distribution with mean $\theta = 1$ can be simulated using `X1 := ExponentialS(1,100):` or using
  `X1 := [seq((-1)*log(1 - rng()),i = 1 .. 100)];`.
- Observations from the distribution with p.d.f. $f(x) = (3/64)x^2$, $0 < x < 4$, and distribution function $F(x) = (1/64)x^3$, $0 < x < 4$, can be simulated using
  `X2 := [seq(4*rng()^(1/3),i = 1 .. 100)];`.
- A random sample from the normal distribution with mean $\mu = 10$ and variance $\sigma^2 = 9$ can be simulated using `X3 := NormalS(10,9,100);`.
- Random numbers from the interval (0,4) can be simulated using
  `X4 := UniformS(0 .. 4,100);`.

**1.3–2** Use data from the exercises in your textbook to construct stem-and-leaf displays and box-and-whisker displays. Interpret your output.

## 1.4 Graphical Comparisons of Data Sets

There are several ways to compare data sets graphically. Perhaps one of the easiest is to draw box plots side by side for two or more sets of data as we did in the last section.

Another graphical technique compares the respective quantiles of the two samples. Order the observations in each set of $n$ observations from independent data sets, say $x_1 \le x_2 \le \cdots \le x_n$ and $y_1 \le y_2 \le \cdots \le y_n$. These are called the order statistics for each of the random samples. Furthermore, each of $x_r$ and $y_r$ is called the quantile of order $r/(n+1)$ or the $100[r/(n+1)]$th percentile.

In a quantile-quantile plot or a $q$–$q$ plot, the quantiles of one sample are plotted against the corresponding quantiles of the other sample. Note that if the samples are exactly the same, the points would all plot on a straight line with slope 1 and intercept 0. The exercises will help illustrate how different means, standard deviations, and shapes affect the $q$–$q$ plot.

When the sample sizes are unequal, the order statistics of the smaller sample are used to determine the quantiles of the larger sample.

A $q$–$q$ plot can also help to indicate whether a particular probability model is appropriate for a particular set of data. To do this, plot the quantiles of the sample against the corresponding quantiles (percentiles) of the distribution. There are procedures that give these percentiles for several standard distributions. A linear $q$–$q$ indicates a good fit. More will be said about this later.

## EXERCISES

**Purpose:** The exercises illustrate the information that can be gained from multiple box plots and $q$–$q$ plots using simulated data, `BoxWhisker`, and `QQ`. This understanding is used to compare sets of data.

**1.4–1** In this exercise random samples of size 100 will be selected from the following distributions:

- Exponential with mean 1, `X1 := ExponentialS(1,100);`,
- $f(x) = (3/64)x^2$, $0 < x < 4$, `X2 := [seq(4*rng()^(1/3),i = 1..100)];`,
- Normal with mean 2 and variance 1, `X3 := NormalS(2,1,100);`,
- Uniform on the interval (0,4), `X4 := UniformS(0 .. 4,100);`.

(a) Use `Percentile` to obtain the 25th, 50th and 75th percentiles of each sample.

(b) Use `BoxWhisker` to make multiple box plots of either pairs of random samples or several random samples. Compare samples from the same or from different distributions. **Experiment!!**

(c) Use `QQ` to make $q$–$q$ plots of pairs of random samples. Analyze the shapes of the $q$–$q$ plots when the samples come from the same distribution or when they come from different distributions. How does reversing the role of the arguments affect the $q$–$q$ plot? Experiment and interpret your output.

**1.4–2** Use `QQ` to obtain $q$–$q$ plots for exercises in your textbook.

**1.4–3** Construct $q$–$q$ plots for random samples from the uniform, normal, and exponential distributions on the $x$-axis with the corresponding percentiles from the theoretical distribution on the $y$-axis.

## 1.5 Time Sequences

In a time sequence observations are recorded in the order in which they were collected as ordered pairs where the $x$-coordinate denotes the time and the $y$-coordinate records the observation. They are useful to denote trends, cycles, or major changes in a process.

### EXERCISES

**Purpose:** The exercises illustrate the information that can be gained from a time sequence.

**1.5–1** Generate 50 random numbers and use `TimePlot` to plot a time sequence of these observations. Do you detect any trends or cycles in your random number generator? Or does the time sequence look random about the line $y = 0.5$?

**1.5–2** Often random errors are normally distributed. To see what a time sequence of random errors could look like, generate 50 observations from a normal distribution with mean 0 and variance 1. Use `TimePlot` to plot a time sequence to see whether the observations are generated in a random order. Use `Histogram` to see whether the histogram of all of the observations tends to be bell shaped.

**1.5–3** Make a time sequence of some of the data in your textbook.

## 1.6 Scatter Plots, Least Squares, and Correlation

Let $(x_1, y_1), (x_2, y_2), \ldots, (x_n, y_n)$ be a set of $n$ data points. The linear least squares regression line is given by

$$\widehat{y} = \widehat{\alpha} + \widehat{\beta}(x - \overline{x})$$

where

$$\widehat{\alpha} = \overline{y} = \frac{1}{n}\sum_{i=1}^{n} y_i$$

and

$$\widehat{\beta} = \frac{\sum_{i=1}^{n}(x_i - \overline{x})(y_i - \overline{y})}{\sum_{i=1}^{n}(x_i - \overline{x})^2} = r\frac{s_y}{s_x}.$$

### EXERCISES

**Purpose:** The exercises give some examples of least squares regression.

**1.6–1** One physical chemistry laboratory experiment involves electron energy transfer from a molecule in the excited electronic state (Benzene) to a different molecule in its ground electronic state (Acetone) and is measured by its fluorescent energy transmission. This experiment is performed to see if it is dependent on the pressure (which is proportional to the concentration) of the acetone. The $x$ values are $P_{Acetone}$ in torr. The $y$ values are $E_t$ in nano amps.

| $x$ | $y$ |
|---|---|
| 0.000 | 101.5 |
| 0.565 | 84.5 |
| 1.093 | 74.9 |
| 1.636 | 64.4 |
| 2.179 | 51.7 |
| 2.707 | 49.5 |
| 3.217 | 46.3 |

Depict a scatter diagram along with the least squares regression line for these $(x, y)$ pairs. Two different uses of the `ScatPlotLine` command are

```
X := [0,0.565,1.093,1.636,2.179,2.707,3.217];
Y := [101.5,84.5,74.9,64.4,51.7,49.5,46.3];
ScatPlotLine(X,Y);
```

and

```
A := [[0,101.5],[0.565,84.5],[1.093,74.9],[1.636,64.4],
[2.179,51.7],[2.707,49.5],[3.217,46.3]];
ScatPlotLine(A);
```

**1.6–2** A study was conducted to observe the effect of exercise on the systolic blood pressure of male subjects 19–20 years of age. The blood pressures of eight subjects were measured before and after several minutes of running up and down the stairs of the science building. Let $x_i$ equal the systolic blood pressure before exercise and let $y_i$ equal the systolic blood pressure after exercise.

| Subject | $x_i$ | $y_i$ |
|---|---|---|
| 1 | 120 | 150 |
| 2 | 119 | 165 |
| 3 | 122 | 177 |
| 4 | 120 | 157 |
| 5 | 121 | 165 |
| 6 | 119 | 167 |
| 7 | 115 | 159 |
| 8 | 122 | 163 |

Use `ScatPlotLine` to depict a scatter diagram along with the least squares regression line.

**1.6–3** Let $X$ and $Y$ denote numbers of students in the library on Monday and Tuesday, respectively. For 9 given times the following numbers were observed:

| Times | 9AM | 11AM | 1PM | 3PM | 5PM | 7PM | 8PM | 9PM | 11PM |
|---|---|---|---|---|---|---|---|---|---|
| $X$ | 23 | 40 | 82 | 79 | 25 | 107 | 128 | 137 | 47 |
| $Y$ | 35 | 36 | 74 | 78 | 20 | 84 | 114 | 97 | 17 |

Use `ScatPlotLine` to make a scatter plot of the 9 points $(x, y)$ with the least squares regression line included on the graph.

**1.6–4** Solve some exercises out of your textbook using `ScatPlotLine`.

# Chapter 2

# Probability

## 2.1 Properties of Probability

Let $A$ be an event associated with a particular experiment. Repeat the experiment $n$ times and let $\#(A)$ denote the number of times that event $A$ occurs in these $n$ trials. An approximation of $p = P(A)$, the probability of event $A$, is the relative frequency of the number of occurrences of $A$. That is,

$$P(A) \approx \frac{\#(A)}{n},$$

the proportion of times that $A$ was observed in $n$ trials.

The probability of the event $A$ is defined by the limit

$$P(A) = \lim_{n \to \infty} \frac{\#(A)}{n}.$$

We let

$$\widehat{p} = \frac{\#(A)}{n}$$

denote the estimate of $p = P(A)$.

If we know $p$, we might be surprised that $\widehat{p}$ is not as close to $p$ as we had expected. It can be shown that if an experiment is repeated $n$ times, 91% of the time $|\,\widehat{p} - p\,| < 1/\sqrt{n}$.

**Remark.** A proof of the above result is given in "The Accuracy of Probability Estimates" by Eugene F. Shuster published in *Mathematics Magazine*, Volume 51, Number 4 (September, 1978), pages 227–229.

Suppose that an experiment has $m$ outcomes, $e_1, e_2, \ldots, e_m$, that are equally likely. If $h$ of these events belong to event $A$, then $P(A) = h/m$.

Let $A$ and $B$ denote identical decks of cards, each deck containing $n$ cards numbered from 1 through $n$. A match occurs if a card numbered $k$, $1 \le k \le n$, occupies the same position in both decks. The probability of at least one match, when using decks containing $n$ cards that have each been shuffled, is given by

$$p_n = 1 - \frac{1}{2!} + \frac{1}{3!} - \frac{1}{4!} + \cdots + \frac{(-1)^{n+1}}{n!}.$$

It is interesting to note that

$$\begin{aligned}\lim_{n\to\infty} p_n &= \lim_{n\to\infty}\left[1 - \left\{1 - \frac{1}{1!} + \frac{1}{2!} - \frac{1}{3!} + \cdots + \frac{(-1)^n}{n!}\right\}\right] \\ &= 1 - e^{-1} \\ &\approx 0.6321.\end{aligned}$$

A problem that is equivalent to the one just stated is the following. Let $P$ denote a permutation of the first $n$ positive integers. A match occurs if integer $k$ occupies the $k$th position in this permutation. The probability of at least one match, in a permutation $P$ that has been selected at random, is given by

$$p_n = 1 - \frac{1}{2!} + \frac{1}{3!} - \frac{1}{4!} + \cdots + \frac{(-1)^{n+1}}{n!}.$$

For example, select the first $n$ positive integers at random, one at a time, without replacement. A match occurs if the integer $k$ is the $k$th integer selected. The probability of at least one match is $p_n$.

Suppose now that $n$ positive integers are selected at random, one at a time, without replacement, out of the first $n$ positive integers. A match occurs if the integer $k$ is the $k$th integer selected. The probability of at least one match is given by

$$q_n = 1 - (1 - 1/n)^n.$$

It is interesting to note that

$$\begin{aligned}\lim_{n\to\infty} q_n &= \lim_{n\to\infty}[1 - (1 - 1/n)^n] \\ &= 1 - e^{-1} \\ &\approx 0.6321.\end{aligned}$$

## EXERCISES

**Purpose:** The relative frequency approach to probability is illustrated by performing some simple experiments involving coins and dice. The random number generator is used to simulate the same experiments so that sample sizes can be quite large. Comparisons of probabilities, $p$, with estimates of $p$ will be made. In addition an area and a "volume" are estimated using the random number generator. The matching problems are illustrated with random experiments and also through simulation on the computer.

**2.1–1** (a) Flip a single coin 20 times.

(b) Calculate the relative frequency of occurrence for each of the events Heads and Tails.

(c) Simulate the experiment of flipping a coin 20 times using random numbers from Appendix C. You could, for example, associate the outcomes [0.00000, 0.49999] and [0.50000, 0.99999] with Tails and Heads on the coin, respectively.

(d) Simulate flipping a coin 250 times. Denote Tails with a zero (0) and Heads with a one (1). This can be done by using one of the following methods:

- `A := Die(2,250):`
  `X := [seq(A[i]-1,i = 1 .. 250)];`
- `X := BernoulliS(0.5,250);` (which will be explained later)
- `X := [seq(floor(rng() + 0.5),i = 1 .. 250)];`

(e) Use `PlotRunningAverage(X);` to check to see if the running averages converge to 0.5.

(f) Calculate the relative frequency of heads $\widehat{p}$ (using `Freq(X,0 .. 1);` may be of help here).

(g) Is your estimate, $\widehat{p}$, within $1/\sqrt{250} = 0.063$ of $p = 1/2$? Compare your results with those of other members of your class.

**2.1–2** (a) Roll a fair 6-sided die (hexahedron) 30 times. Record the number of occurrences and the relative frequency of each of the 6 possible outcomes.

(b) Use a calculator random number generator or random numbers from Appendix C to simulate the roll of a die 30 times. Record the number of occurrences and the relative frequency of each of the 6 possible outcomes.

(c) Use `X := Die(6,300);` to simulate 300 trials of this experiment. Calculate the relative frequency of each of the 6 possible outcomes (use `Freq(X,1 .. 6)`).

(d) Let $\widehat{p}_2$ denote the proportion of observed twos. Is $\widehat{p}_2$ within $1/\sqrt{300} = 0.058$ of $p_2 = 1/6 = 0.167$? Compare your results with other members of your class.

(e) Use `Histogram(X,0.5 .. 6.5,6);` to plot a relative frequency histogram of your data.

**2.1–3** Use a random number to simulate the experiment of dividing a line segment of length one into two parts. Let $A$ be the event that the larger segment is at least twice as long as the shorter segment.

(a) Calculate $P(A)$, the probability of event $A$.

(b) Simulate this experiment 100 times using `X := RNG(100):` and compare the relative frequency of $A$ with $P(A)$ (it may help to use `sort`). Or use the following:

```
X := RNG(100):
num := 0;
for k to 100 do
if 1/6 < abs(X[k] - 0.5) then num := num+1 fi
od:
num/100;
```

**2.1–4** Let $A = \{(x_1, x_2) : x_1^2 + x_2^2 \leq 1\}$. Let $B = \{(x_1, x_2) : 0 \leq x_1 < 1, 0 \leq x_2 < 1\}$.

(a) Sketch the sets $A$, $B$, and $C = A \cap B$.

(b) If $X_1$ and $X_2$ denote a pair of random numbers, is it clear that the probability that the pair of random numbers falls in C is equal to $\pi/4$? That is, $P[(X_1, X_2) \in C] = \pi/4$? Explain.

(c) Generate $m = 1500$ pairs of random numbers. Find the value of

$$Y = \#(\{(x_1, x_2) : x_1^2 + x_2^2 \leq 1\}),$$

the number of pairs of random numbers that fall in $C = A \cap B$. Note that this problem is continued in the next exercise. **Hint:** Here are two possible approaches.

```
X := RNG(1500):
Y := RNG(1500):
C := [seq(X[i]^2 + Y[i]^2,i = 1 .. 1500)]:
SC := sort(C):
```

Now determine $Y$ by inspection through such commands as `SC[1165 .. 1200];`.

```
randomize();
n := 2;
m := 1500;
Y := 0:
for k from 1 to m do
s := 0;
for j from 1 to n do  s := s+rng()^2 od;
if s < 1 then Y := Y+1 fi
od:
Y;
```

(d) Is $Y/1500$ within $1/\sqrt{1500} = 0.0224$ of $\pi/4 = 0.7854$?

(e) An estimate of $\pi$, the area of $A$, is given by $2^2(Y/1500)$. Give this estimate of $\pi$ using the results of your simulation.

**2.1–5** Let $B_n = \{(x_1, x_2, \ldots, x_n) : x_1^2 + x_2^2 + \cdots + x_n^2 \le 1\}$ denote a ball of radius one in $E^n$, Euclidean $n$-space. Let $V_n$ denote the "volume" of $B_n$.

(a) What are the values of $V_1$, $V_2$, and $V_3$?

(b) Estimate $V_4$, $V_5$, and $V_6$, using simulation techniques similar to those in Exercise 2.1–4, parts (c) and (e).

(c) For what value of $n$ is $V_n$ a maximum?

**2.1–6** Take two "identical" decks of cards, each deck containing $n$ cards. (For example, from a standard deck of cards, use the 13 spades as one deck and the 13 hearts as the other deck.) Shuffle each of the decks and determine if at least one match has occurred by comparing the positions of the cards in the decks. If a match occurs at one or more than one position, call the *trial* a success. Thus, each experiment consists of comparing the positions of the $n$ cards in each of the two decks. The outcome of the experiment is "success" if at least one match occurs, "failure" if no matches occur.

(a) Repeat this experiment 10 times for each of $n$ = 4, 6, 8, 10, and 12. Record your results in Table 2.1–1.

| $n$ | Number of Successes | Number of Failures | Relative Frequency of Success for the Class | Probability of Success |
|---|---|---|---|---|
| 4 | | | | |
| 6 | | | | |
| 8 | | | | |
| 10 | | | | |
| 12 | | | | |

**Table 2.1–1**

(b*) Simulate this experiment on the computer using permutations of the first $n$ positive integers. For example, for $n = 6$, you could randomly order $1, 2, 3, 4, 5, 6$ (by using `randperm(6)`) and if any of these integers occupies its natural position, a match would occur. Repeat the simulation 100 times and calculate the relative frequency of success. (See Exercise 3.7-12.) **Hint:**

```
with(combinat,randperm);
randomize();
```

```
n := 6;
m := 100;
L := [seq(randperm(n),i = 1 .. m)]:
for k from 1 to m do
XX[k] := 0:
for j from 1 to n do
if L[k][j] = j then XX[k] := XX[k]+1 fi
od
od:
X := [seq(XX[i],i = 1 .. m)];
Freq(X,0 .. n);
```

**2.1–7** Roll an $n$-sided die $n$ times. A match occurs if side $k$ is the outcome on the $k$th roll. The experiment is a "success" if at least one match occurs, a "failure" if no matches occur. (Note that as soon as a match has occurred, the experiment can be terminated.)

(a) Repeat this experiment 10 times for each of $n = 4$ (tetrahedron), 6 (hexahedron), 8 (octahedron), and 12 (dodecahedron). An icosahedron (20 sides) can be used for 10 equally likely integers. Record the results of your experiment in Table 2.1–2.

| $n$ | Number of Successes | Number of Failures | Relative Frequency of Success for the Class | Probability of Success |
|---|---|---|---|---|
| 4 | | | | |
| 6 | | | | |
| 8 | | | | |
| 10 | | | | |
| 12 | | | | |

**Table 2.1–2**

(b) Simulate this experiment on the computer. Repeat the simulation 100 times for each of $n$ = 4, 6, 8, 10, and 12. Calculate each relative frequency of success, comparing the relative frequencies with their respective probabilities. **Hint:** In Exercise 2.1-6, let `L := [seq(Die(n,n),i = 1 .. m)]:`.

**2.1–8** Construct a table that gives the values of $p_n$ and $q_n$ as defined in this section for $n = 1$ to 15. Comment on the relative rates of convergence of $p_n$ and $q_n$ to $1 - e^{-1}$.
**Hint:** Values such as $q_5, q_{10}, q_{15}, \ldots, q_{100}$ can be obtained by using

```
q := <1 - (1 - 1/n)^n|n>;
[seq(evalf(q(5*i)),i = 1 .. 20)];
```

## Questions and Comments

**2.1–1** In Exercise 2.1–4(c), $Y$ is a random variable that has a binomial distribution (see Section 3.5). Most of the time (about 95% of the time) the observed value of $Y$ will lie in the interval [1146, 1209].

**2.1–2** For Exercise 2.1–5, verify that $V_{2n} = \pi^n/n!$ and $V_{2n-1} = 2^n\pi^{n-1}/(1 \cdot 3 \cdots (2n-1))$ for $n = 1, 2, 3, \ldots$. Because $V_5 - V_6 = 0.096076$, a large sample is required to "verify," using simulation techniques, that the maximum is $V_5$.

**Remark.** A discussion of volumes of balls in $n$-space is given in "Volume of An $n$-dimensional Unit Sphere" by Ravi Salgia, published in *Pi Mu Epsilon Journal*, Volume 7, Number 8, (Spring 1983), pages 496–501.

**2.1–3** A more general formula for the volume of a ball of radius $r > 0$ in $n$-space is

$$V_n(r) = \frac{r^n \pi^{n/2}}{\Gamma\left(\frac{n}{2} + 1\right)}, \quad n = 0, 1, 2, \ldots.$$

Use this formula to show

$$\sum_{n=0}^{\infty} V_{2n} = e^{\pi}.$$

**Remark.** See "How Small Is a Unit Ball?" by David J. Smith and Mavina K. Vamanamurthy in *Mathematics Magazine*, Vol. 62, No. 2, April 1989, pages 101–107.

**2.1–4** Prove that the probabilities given for $p_n$ in Exercise 2.1–6 are correct. For example, let $A_i$ denote a match on the $i$th draw. For $n = 4$, find $P(A_1 \cup A_2 \cup A_3 \cup A_4)$ by a natural extension of the probability of the union of two or three events. Note that $P(A_i) = 3!/4!$, $P(A_i \cap A_j) = 2!/4!$, $P(A_i \cap A_j \cap A_k) = 1/4!$.

**2.1–5** Prove that the probabilities given for $q_n$ in Exercise 2.1–7 are correct. Let $B_k$ denote the event that integer $k$ is the $k$th integer selected. Then $P(B_k) = 1/n$, $P(B_k \cap B_j) = (1/n)^2$, etc. Now find $P(B_1 \cup B_2 \cup \cdots \cup B_n)$ by a natural extension of the probability of the union of two or three events.

**2.1–6** For an article on matching probabilities, see "The Answer is $1 - 1/e$. What is the Question?" by Elliot A. Tanis, published in *Pi Mu Epsilon Journal*, Volume 8, Number 6, (Spring 1987), pages 387–389.

## 2.2 Methods of Enumeration

The number of permutations of a set of $n$ objects is $n(n-1)\cdots(2)(1) = n!$. The number of permutations of a set of $n$ objects taken $r$ at a time is

$$n(n-1)\cdots(n-r+1) = \frac{n!}{(n-r)!} = P(n,r).$$

The number of distinguishable permutations of a set of $n$ objects, $r$ of one kind and $n-r$ of another kind, is

$$\frac{n!}{r!(n-r)!} = \binom{n}{r} = C(n,r).$$

The number of ordered samples of size $r$, sampling with replacement from a set of $n$ objects, is $n^r$.

The number of ordered samples of size $r$, sampling without replacement from a set of $n$ objects, is

$$n(n-1)\cdots(n-r+1) = \frac{n!}{(n-r)!} = P(n,r).$$

The number of unordered samples of size $r$, sampling without replacement from a set of $n$ objects, or the number of combinations of a set of $n$ objects taken $r$ at a time, is

$$\frac{n!}{r!(n-r)!} = \binom{n}{r} = C(n,r).$$

## EXERCISES

**Purpose:** By listing and counting permutations and combinations of sets of objects, the exercises illustrate relations between permutations and combinations, between ordered and unordered samples. The exercises also illustrate the difference between sampling with and without replacement. Before starting, use the *Maple* `help` facility to become familiar with the `choose`, `permute`, `randcomb`, and `randperm` commands.

**2.2–1** As $n$ increases, $n!$ increases very fast. In particular, for large $n$, $n! > n^{10}$ (in fact, $n!$ is larger than any power of $n$ for large enough $n$). Show this by plotting $n!$ and $n^{10}$ on one set of axes. Approximately how large must $n$ be for $n!$ to be larger than $n^{10}$? How large must $n$ be for $n!$ to be larger than $10^n$? (The latter is larger than $n^{10}$.)

**2.2–2** Stirling's formula gives an approximation for $n!$ through $n^n e^{-n}\sqrt{2\pi n}$. Compare $n!$ to its Stirling approximation when $n = 5$, 10, 25, and 50. How good is this approximation for large $n$?

**2.2–3** (a) What is the number of permutations of three objects?

(b) List all of the permutations of the three words Red, White, and Blue.

(c) What is the number of permutations of four objects taken three at a time?

(d) List all of the permutations of the four words, Red, White, Blue, and Yellow, taken three at a time.

```
with(combinat,permute);
permute([Red,White,Blue]);
permute([Red,White,Blue,Yellow],3);
```

**2.2–4** (a) What is the number of permutations of the five words, Red, White, Blue, Green, and Yellow, taken three at a time?

(b) Obtain a listing of these permutations. Enter

```
with(combinat,permute);
L := [Red,White,Blue,Green,Yellow];
permute(L,3);
```

**2.2–5** (a) What is the number of ordered samples of size 3, sampling without replacement, that can be taken from a set of 5 objects?

(b) If the 5 objects are the words, Red, White, Blue, Green, and Yellow, obtain a listing of the $P(5,3)$ ordered samples using `permute`.

(c) What is true about your answers to Exercises 2.2–4(b) and 2.2–5(b)?

**2.2–6** (a) What is the number of unordered samples of size 3, sampling without replacement, that can be taken from a set of 5 objects?

(b) What is the number of combinations of 5 objects taken 3 at a time?

(c) If the 5 objects are the words Red, White, Blue, Green, and Yellow, use `choose` to obtain a listing of the combinations of these 5 objects taken 3 at a time.

```
with(combinat,choose);
L := [Red,White,Blue,Green,Yellow];
choose(L,3);
```

**2.2–7** (a) Through *Maple*, obtain a listing of the combinations of the five words Paint, Box, The, Violet, and You, taken 4 at a time.

(b) Obtain a listing of the permutations of the five words of part (a) taken 4 at a time.

**2.2–8** (a) Obtain a listing of all of the permutations of the four words: Paint, Box, The, Violet.

(b) How many of the permutations in part (a) "make sense" as a sentence?

**2.2–9** In *Maple*, $\binom{n}{k}$ is designated by `binomial(n,k);`. For a specific $n$, say $n = 20$, $\binom{n}{k}$ can be plotted as a function of $k$. To get a sense of the "shape" of $\binom{n}{k}$ plot $\binom{n}{k}$ as a function of $k$ for several choices of $n$. Obtain a 3-dimensional plot of $\binom{n}{k}$ for $20 \le n \le 25$ and $5 \le k \le 20$. Approximately how large does $\binom{n}{k}$ get for these values of $n$ and $k$? For a fixed $n$, what value of $k$ seems to make $\binom{n}{k}$ largest? You may use

```
plot3d(binomial(n,k),n = 20 .. 25,k = 2 .. 20,axes=BOXED);
```

## Questions and Comments

**2.2–1** Let $w$ denote the number of unordered samples of size 3 that can be taken from a set of 5 objects. Each of these $w$ samples can be ordered in $3! = 6$ ways. This sequence of operations yields $(w)(3!) = 5!/(5-3)!$ ordered samples of size 3 taken from 5 objects.

**2.2–2** If the topics of permutations and combinations and their relations to each other are still not clear, write and work additional problems that use `choose` and `permute`.

## 2.3 Conditional Probability

Let $A$ and $B$ be two events. The conditional probability of event $B$, given that event $A$ has occurred, is defined by

$$P(B \mid A) = \frac{P(A \cap B)}{P(A)}, \text{ provided } P(A) > 0.$$

Conditional probabilities can be used to solve the "birthday problem" and a variety of "matching problems." Select at random a single card from each of six decks of playing cards. A match occurs if two of the six cards selected are identical. The probability that all six cards are different is

$$q = \frac{52}{52} \cdot \frac{51}{52} \cdot \frac{50}{52} \cdot \frac{49}{52} \cdot \frac{48}{52} \cdot \frac{47}{52} = 0.741.$$

Thus the probability of at least one match is

$$p = 1 - q = 0.259.$$

## EXERCISES

**Purpose:** The meaning of conditional probability is illustrated empirically.

**2.3–1** An urn contains eight white balls and two red balls. The balls are drawn from the urn one at a time, sampling without replacement.

(a) What is the probability of drawing a red ball on the (i) first draw, (ii) second draw, (iii) sixth draw?

(b) Perform this experiment 10 times. Record your outcomes in Table 2.3–1.

| Red On | For Me | | For Class | | Relative Frequency Of Yeses for Class | Probability Of Red |
|---|---|---|---|---|---|---|
| | Yes | No | Yes | No | | |
| 1st Draw | | | | | | |
| 2nd Draw | | | | | | |
| 6th Draw | | | | | | |

**Table 2.3–1**

**2.3–2** Urn $A$ contains 1 red and 1 white ball, urn $B$ contains 2 red balls, urn $C$ contains 2 white balls. An urn is selected at random and a ball is drawn from the urn.

(a) What is the probability of drawing a red ball?

(b) If a red ball has been drawn, what is the probability that it was drawn from urn $A$?

(c) If a red ball has been drawn, what is the probability that the other ball in the urn is red? That is, what is the probability that you are drawing from urn $B$?

(d) Check your answers empirically. Either perform the experiment with balls and urns or simulate this experiment on the computer.

**2.3–3** (Craps) You, the player, roll a pair of dice and observe the sum. If you roll a 7 or 11, you win. If you roll a 2, 3, or 12, you lose. If you roll a 4, 5, 6, 8, 9, or 10 on the first roll, this number is called your point. You now continue to roll the dice until either your point is rolled, in which case you win, or a 7 is rolled, in which case you lose.

(a) Show that the probability of rolling an 8 on your first roll and then winning is equal to $(5/36)(5/11)$.

(b) Show that the total probability of winning at craps is $244/495 = 0.492929\ldots$.

(c) `Craps();` simulates a single craps game. The output of `Craps();` is a list with its first entry equal to 0 or 1, depending on whether the player wins or looses the game, and the remaining entries are the outcomes of the rolls of the die. For example, `[0, 5, 6, 6, 8, 7]` indicates that the player lost on successive rolls of 5, 6, 6, 8, and 7. Use `Craps();` to simulate several craps games.

(d) `FastCraps(n);` simulates `n` craps games and returns a list of 0s and 1s indicating wins and losses (the exact sequence of die rolls is suppressed). Verify your answer to part (b) using simulation. In particular simulate $n = 500$ trials of this game and use `PlotRunningAverage` to see if the average number of wins converges. Also, count the number of trials on which there is a win. (See part (e).) Compare the relative frequency of wins with 0.49293.

(e*) In craps it is possible to win or lose with just one roll of the dice. At times it takes several rolls of the dice to determine whether a player wins or loses. Let $X$ equal the number of times the dice are rolled to determine whether a player wins or loses. Use 500 simulations to estimate $P(X = k)$, $k = 1, 2, 3, 4, \ldots$. Show that the average of the observations of $X$ is close to $E(X) = 3.38$. Show that the sample variance of the observations of $X$ is close to $\text{Var}(X) = 9.02$. **Hint:**

```
randomize();
for k from 1 to 500 do
Y := Craps(); XX[k] := nops(Y)-1
od:
X := [seq(XX[k],k = 1 .. 500)]:
Mean(X);
Variance(X);
```

**Remark.** A theoretical discussion of part (d) is given in an article by Armond V. Smith, Jr., "Some Probability Problems in the Game of 'Craps'," *The American Statistician*, June, 1968, pages 29–30.

**2.3–4** A set of cards is made up of the four queens and four kings from a standard deck of playing cards (in *Maple*, you can let `C` represent these cards by `C:=[CQ, CK, DQ, DK, HQ, HK, SQ, SK];`). Two cards are drawn from this deck at random.

(a) Use `choose(C,2);` to obtain a listing of the $\binom{8}{2} = 28$ possible sets of two cards.

(b) Consider the following events: $A = \{\text{Two Queens}\}$, $B = \{\text{at least one card is a Queen}\}$, $C = \{\text{at least one card is a black Queen}\}$, $D = \{\text{one card is the Queen of Spades}\}$. Find, by counting outcomes in part (a), the number of outcomes that are in each of these events, namely, $N(A)$, $N(B)$, $N(C)$, $N(D)$.

(c) Compute the following probabilities: $P(A)$, $P(A \mid B)$, $P(A \mid C)$, $P(A \mid D)$.

**2.3–5** Using six decks of playing cards, shuffle each of the decks and then draw one card at random from each of the decks. If at least one match occurs, call the trial a success.

(a) Repeat this experiment 20 times. What proportion of the 20 trials resulted in success?

(b) Simulate this experiment 100 times on the computer by generating 100 sets of 6 integers selected at random from the first 52 positive integers, sampling with replacement. What proportion of the 100 trials resulted in success? (Recall that `D:=Die(52,6);` simulates six rolls of a "52-sided die" and `L:=Freq(D,1 .. 52);` will show where matches occur.)

## Questions and Comments

**2.3–1** If the answers to Exercise 2.3–4(c) do not make sense to you, simulate this experiment several times. Compare relative frequencies of events with probabilities.

# 2.4 Independent Events

The events $A$ and $B$ are independent if $P(A \cap B) = P(A)P(B)$, in which case $P(B|A) = P(B)$, provided $P(A) > 0$. If two events are not independent, they are called dependent events.

## EXERCISES

**Purpose:** The exercises illustrate empirically the ideas of independent and dependent events.

**2.4–1** A red die and a white die are rolled. Let event $A = \{4 \text{ on red die}\}$ and event $B = \{\text{sum of dice is odd}\}$.

(a) Prove that events $A$ and $B$ are independent.

(b) Simulate 100 rolls of a pair of dice, one red and the other white. For each trial print the outcome on the red die, the outcome on the white die, and the sum of the two dice. (See Exercise 2.4–2(b).)

(c) The independence of $A$ and $B$ means empirically that

$$\frac{\#(A \cap B)}{100} \approx \frac{\#(A)}{100} \cdot \frac{\#(B)}{100},$$

$$\frac{\#(A \cap B)}{\#(A)} \approx \frac{\#(B)}{100},$$

$$\frac{\#(A \cap B)}{\#(B)} \approx \frac{\#(A)}{100}.$$

For your set of data determine whether the above three approximate equalities hold.

**2.4–2** A red die and a white die are rolled. Let event $C = \{\text{sum of dice is } 11\}$ and event $D = \{5 \text{ on red die}\}$.

(a) Prove that events $C$ and $D$ are dependent.

(b) Use the data generated in Exercise 2.4–1(b) to illustrate empirically that $P(D|C) \neq P(D)$ . That is, check whether

$$\frac{\#(C \cap D)}{\#(C)} \approx \frac{\#(D)}{100}.$$

(c) Check where the following approximate equalities hold:

$$\frac{\#(C \cap D)}{\#(D)} \approx \frac{\#(C)}{100},$$

$$\frac{\#(C)}{100}\frac{\#(D)}{100} \approx \frac{\#(C \cap D)}{100}.$$

Should you have expected them to hold?

**2.4–3** A rocket has a built-in redundant system. In this system, if component $K_1$ fails, it is by-passed and component $K_2$ is used. If component $K_2$ fails, it is by-passed and component $K_3$ is used. The probability of failure of any one of these three components is $p = 0.40$ and the failures of these components are mutually independent events.

(a) What is the probability that this redundant system does not fail?

(b) Simulate 100 flights of this rocket. For each flight determine whether or not component $K_1$ fails. If $K_1$ fails, determine whether or not component $K_2$ fails. If $K_2$ fails, determine whether or not component $K_3$ fails.

(c) Calculate the proportion of time that the system does not fail. Is this proportion close to your answer in part (a)?

(d) Repeat this problem using a different value for $p$, the probability of failure of a component. For what value of $p$ would you be willing to use this rocket to launch a person into space?

## Questions and Comments

**2.4–1** In Exercise 2.4–1(c), explain why the three approximate equalities should hold.

## 2.5 Bayes' Formula

Let the events $B_1, B_2, \ldots, B_m$ constitute a partition of the sample space $S$. Let $A \subset S$. If $P(A) > 0$, then

$$P(B_k \mid A) = \frac{P(B_k)P(A \mid B_k)}{\sum_{i=1}^{m} P(B_i)P(A \mid B_i)}, \quad i = 1, 2, \ldots, m.$$

This is Bayes' Formula.

## EXERCISES

**Purpose:** Bayes' Formula is illustrated empirically using the data in Table 2.5–3 which depends on the data in Table 2.5–1.

**2.5–1** Urn $B_1$ contains three white balls and six red balls, urn $B_2$ contains six white balls and three red balls, urn $B_3$ contains four white balls and five red balls. Roll a fair die. If the outcome is 1 or 2, select a ball from urn $B_1$, if the outcome is 3, select a ball from urn $B_2$, if the outcome is 4, 5, or 6, select a ball from urn $B_3$.

(a) What is the probability of drawing a red ball? A white ball?

(b) Perform this experiment 20 times. Record your outcomes in Table 2.5–1 and summarizing Table 2.5–2.

| Trial | 1 | 2 | 3 | 4 | 5 | 6 | 7 | 8 | 9 | 10 |
|---|---|---|---|---|---|---|---|---|---|---|
| Urn | | | | | | | | | | |
| Color | | | | | | | | | | |
| Trial | 11 | 12 | 13 | 14 | 15 | 16 | 17 | 18 | 19 | 20 |
| Urn | | | | | | | | | | |
| Color | | | | | | | | | | |

**Table 2.5–1**

| | Number of Occurrences | | Relative Frequency | | |
|---|---|---|---|---|---|
| | For Me | For Class | For Me | For Class | Probability |
| Red | | | | | |
| White | | | | | |
| Total | | | | | |

**Table 2.5–2**

(c) Given that the drawn ball is red, what is the conditional probability that it was drawn from urn $B_1$? That is, what is the value of $P(B_1 \mid R)$?

(d) Considering only those outcomes in part (b) that yielded a red ball, record the results in Table 2.5–3.

| | Number of Occurrences | | Relative Frequency | | |
|---|---|---|---|---|---|
| | For Me | For Class | For Me | For Class | $P(B_1 \mid R)$ |
| Urn $B_1$ | | | | | |
| Urn $B_2$ | | | | | |
| Urn $B_3$ | | | | | |

**Table 2.5–3**

**2.5–2** Simulate 100 trials of the experiment in Exercise 2.5–1 on the computer, obtaining results like those asked in parts (b) and (d) of that exercise.

**2.5–3** On a TV game show there are three closed doors. Suppose that there is a new car behind one of the doors and a goat behind each of the other two doors. A contestant selects a door at random and wins the prize behind it unless he or she wants to switch doors as now described. Before opening the contestant's door to reveal the prize, here are three possible rules that the host could use to decide whether to open the door:

(i) The host of the show always opens one of the other two doors and reveals a goat, selecting the door randomly when each of these doors hides a goat.
(ii) The host only opens one of the other two doors to reveal a goat, selecting the door randomly, when the contestant has selected the door with a car behind it.
(iii) When the contestant's door hides a car, the host randomly selects one of the other two doors and opens it to reveal a goat. When the contestant's door hides a goat, half the time the host opens the other door that is hiding a goat.

The contestant is allowed to switch doors before the contestant's door is opened to reveal their prize. Suppose that the contestant initially selects door 1. For each of the three strategies by the host, find the probability that the contestant wins a car (a) without switching, (b) with switching. Write a computer program to confirm your answers using simulations.

# Chapter 3

# Discrete Distributions

## 3.1 Random Variables of the Discrete Type

Let $X$ be a random variable of the discrete type. Let $R$ be the support of $X$. The probability density function (hereafter p.d.f.) of $X$ is defined by

$$f(x) = P(X = x), \quad x \in R \quad \text{(zero elsewhere)}.$$

The distribution function of $X$ is defined by

$$F(x) = P(X \leq x), \quad -\infty < x < \infty.$$

The graph of the p.d.f. of $X$ is a plot of the points $\{(x, f(x)) : x \in R\}$. Another way to visualize the graph of the p.d.f. is with a probability histogram. Suppose that $R = \{0, 1, 2, \ldots, n\}$. If $k \in R$, let $g(x) = f(k)$, $k - 1/2 < x \leq k + 1/2$, (zero elsewhere). The graph of $y = g(x)$, $-1/2 < x \leq n + 1/2$, is called a probability histogram. (Obvious modifications are made in this definition for different sets $R$.) Note that the total area under $y = g(x)$ is equal to one.

### EXERCISES

**Purpose:** These exercises use the functions `ProbHist` and `PlotDiscCDF` for graphing probability histograms and distribution functions for discrete random variables.

**3.1–1** Let $f(x) = x/10$, $x = 1, 2, 3, 4$, be the p.d.f. of the discrete random variable $X$.

(a) Use `ProbHist` to depict a probability histogram for this distribution.

(b) Use `PlotDiscCDF` to plot the distribution function of $X$.

```
y := x/10;
ProbHist(y,1 .. 4);
PlotDiscCDF(y,1 .. 4);
```

**3.1–2** Let $f(x) = 1/10$, $x = 0, 1, 2, \ldots, 9$, be the p.d.f. of the discrete random variable $X$.

(a) Use `ProbHist` to draw a probability histogram of this p.d.f.

(b) Use `PlotDiscCDF` to draw the distribution function of $X$.

(c) Simulate 200 observations of $X$ using `DiscUniformS`. Or you could use `X := Die(10,200):`, subtracting 1 from each observation and storing the resulting data in a list. Use these observations and `ProbHist` and `Histogram` to plot a relative frequency histogram with the probability histogram superimposed. Now use `PlotDiscCDF` and `PlotEmpCDF` to obtain the empirical distribution function with the theoretical distribution function, respectively. Produce a superimposed graph of these two.

```
X := DiscUniformS(0 .. 9,200);
y := 1/10;
p1 := ProbHist(y,0 .. 9):
p2 := Histogram(X,-0.5 .. 9.5,10).
plot({p1,p2});
p3 := PlotDiscCDF(y,0 .. 9):
p4 := PlotEmpCDF(X,0 .. 9):
plot({p3,p4});
```

**3.1–3** Let $f(x) = (5 - x)/10$, $x = 1, 2, 3, 4$, be the p.d.f. of the discrete random variable $X$.

(a) Use `ProbHist` to depict a probability histogram.

(b) Use `PlotDiscCDF` to depict the distribution function of $X$.

**3.1–4** Let $f(x) = (\frac{1}{4})(\frac{3}{4})^{x-1}$, $x = 1, 2, 3, \ldots$, be the p.d.f. of the discrete random variable $X$.

(a) Use `ProbHist` to depict the probability histogram for $x = 1, 2, \ldots, 16$, only.

(b) Use `PlotDiscCDF` to depict the distribution function of $X$ for $x$ between 1 and 16, inclusive.

**3.1–5** Let $X$ equal the sum when a pair of 4-sided dice is rolled. The p.d.f. of $X$ is given by $f(x) = (4 - \mid x - 5 \mid)/16$ for $x = 2, 3, 4, 5, 6, 7, 8$.

(a) Use `Die` (twice) to simulate 200 observations of $X$.

(b) Plot a relative frequency histogram with the p.d.f. superimposed.

```
X1 := Die(4,200):
X2 := Die(4,200):
X := [seq(X1[i]+X2[i],i = 1 .. 200)];
```

```
y := 1/4-1/16*abs(x-5);
p1 := ProbHist(y,2 .. 8):
p2 := Histogram(X,1.5 .. 8.5,7):
plot({p1,p2});
```

(c) Use the observations of $X$, `PlotEmpCDF`, and `PlotDiscCDF` to plot the empirical distribution function with the theoretical distribution function superimposed.

## Questions and Comments

**3.1–1** The comparison of empirical data and a theoretical model, like those in Exercises 3.1–2 and 3.1–5, will be illustrated frequently. Be sure that you understand this concept and how it is illustrated graphically.

# 3.2 Mathematical Expectation

If $f(x)$ is the p.d.f. of the random variable $X$ of the discrete type with space $R$, then

$$E[u(X)] = \sum_{x \in R} u(x) f(x)$$

is called the mathematical expectation or expected value of the function $u(X)$, provided this sum exists.

## EXERCISES

**Purpose:** These exercises illustrate the fact that expectation is a weighted average.

**3.2–1** A bowl contains 5 chips. Three chips are numbered 2 and two chips are numbered 4. Draw one chip at random from the bowl and let $X$ equal the number on the chip.

(a) Define the p.d.f. of $X$, $f(2) = 3/5, f(4) = 2/5$, with a *Maple* procedure.

(b) If a payoff of $k$ dollars is given for drawing a chip numbered $k$, for $k = 2$ or $k = 4$, what is the expected value of payment? That is, find $E(X) = \sum_{x \in R} x f(x), R = \{2, 4\}$ using a *Maple* command.

(c) Suppose that the payment is equal to the square of the number on the selected chip. What then is the expected value of payment? That is, find $E(X^2)$.

(d) Suppose that the payment is equal to the cube of the number on the selected chip. What then is the expected value of payment? That is, find $E(X^3)$.

```
f := proc(x)
if x = 2 then 3/5
elif x = 4 then 2/5
else 0
fi
end;
sum('x*f(x)',x = 2 .. 4);
sum('x^2*f(x)',x = 2 .. 4);
sum('x^3*f(x)',x = 2 .. 4);
```

(e) Use `DiscreteS` to simulate 100 trials of this experiment. Find the averages of the 100 outcomes, the squares of the 100 outcomes, and the cubes of the 100 outcomes. Compare these three averages with $E(X)$, $E(X^2)$, and $E(X^3)$, respectively.

```
randomize();
L := [2,3/5,4,2/5];
X := DiscreteS(L,100):
Freq(X,2 .. 4);
Mean(X);
X2 := [seq(X[i]^2,i = 1 .. 100)]:
X3 := [seq(X[i]^3,i = 1 .. 100)]:
evalf(Mean(X2));
evalf(Mean(X3));
```

**3.2–2** Let $X$ equal an integer selected at random from the first $n$ positive integers. The p.d.f. of $X$ is $f(x) = 1/n,\ x = 1, 2, \ldots, n$.

(a) If a payoff of $1/x$ dollars is given for selecting integer $x$, $x = 1, 2, \ldots, n$, find the expected value of payment. That is, compute

$$E(1/X) = \sum_{x=1}^{n}(1/x)f(x)$$

when $n = 5$ and when $n = 10$. You can use the *Maple* `sum` function to compute $E(1/X)$ (`n := 10; sum(1/n/x,x = 1 .. n);`).

(b) Use `Die(10,200)` to simulate 200 repetitions of this experiment for $n = 10$. Calculate the average of the 200 payoffs, $1/x$. Compare this average with the value of E($1/X$) computed in part (a). Try

```
L := Die(10,200);
Pay := [seq(1/L[i],i = 1 .. 200)];
Mean(Pay);
```

or

```
L := Die(10,200);
(1/200)*sum(1/L[i],i = 1 .. 200);
```

(c) Define $E(1/X)$ as a *Maple* expression or function of $n$ and study its behavior as $n$ gets large (e.g., consider whether it is monotone or has a limit). Explain why $E(1/X)$ behaves the way that it does. To define $E(1/X)$ as a function, you may use

```
g := n -> sum(1/x/n,x = 1 .. n);
```

**3.2–3** Three unbiased dice are rolled. Let $X$ equal the observed number of fours. The p.d.f. of $X$ is $f(0) = 125/216, f(1) = 75/216, f(2) = 15/216$, and $f(3) = 1/216$.

(a) The payoff for this game is equal to \$1 for each four that is rolled. It costs \$1 to play this game and this dollar is lost if no fours are rolled. The dollar is returned to the player if 1, 2, or 3 fours are rolled. Show that the expected payoff, or expected value of this game, is equal to $-17/216 = -7.87$ cents. (This problem is based on the game of chance called chuck-a-luck.) Graph a probability histogram of the payoffs. Use either

```
P := [-1,125/216,1,25/72,2,5/72,3,1/216];
mu := sum(P[2*i-1]*P[2*i],i = 1 .. 4);
ProbHist(P);
```

or

```
f := proc(x)
if x = -1 then 125/216
elif x = 1 then 25/72
elif x = 2 then 5/72
elif x = 3 then 1/216
else 0
fi
end;
mu := sum('x*f(x)',x = -1 .. 3);
ProbHist('f(x)',-1 .. 3);
```

(b) Illustrate empirically that the expected value of this game is $-7.87$ cents. In particular simulate 500 plays of this game and find the average value of the payoffs. Note that the possible values of the payoffs are $-1$, 1, 2, and 3. Use either

```
X := DiscreteS(P,500):
Mean(X);
```

or

```
randomize();
for k from 1 to 500 do
XX[k] := 0;
D := Die(6,3);
for j to 3 do
if D[j] = 4 then XX[k] := XX[k]+1 fi
od;
if XX[k] = 0 then XX[k] := -1 fi
od:
X := [seq(XX[k],k = 1 .. 500)]:
Mean(X);
```

(c) Graph the probability histogram and the histogram of the data on the same graph.

(d) Suppose that you are given 5 dollars to play this game and that you place a single bet of 1 dollar per minute. Let $t$ denote the time required to lose this 5 dollars. Find $E(T)$.

(e*) Investigate the behavior of $T$ by simulating 200 observations of $T$. Is the average of these 200 observations close to $E(T)$? Are the observations of $T$ close to each other or are they spread out? The variance (or standard deviation) gives a measure of the spread of the distribution. Use `PlotRunningAverage(T);` and `BoxWhisker(T);` to investigate the behavior of the observations of $T$.

```
for k from 1 to 200 do
CAP := 5;
TT[k] := 0:
while 0 < CAP do
win := 0;
D := Die(6,3);
for j to 3 do
if D[j] = 4 then win := win+1 fi
od:
if win = 0 then win := -1 fi:
TT[k] := TT[k]+1;
CAP := CAP+win
od
od:
T := [seq(TT[k],k = 1 .. 200)];
Mean(T);
```

**3.2–4** In a particular lottery a two digit integer is selected at random between 00 and 99, inclusive. A one dollar bet is made that a particular integer will be selected. If that integer is selected the payoff is 50 dollars. Win or lose, the one dollar is not returned to the bettor. Let $X$ denote the payoff to the bettor.

(a) Calculate $E(X)$.

(b) Simulate 1000 observations of $X$. The number that is bet on may be the same for all of the trials or you may pick the number using some scheme. For each trial note that $X = -1$ or $X = 49$. Is the average of the 1000 observations of $X$ close to your answer in part (a)?

## Questions and Comments

**3.2–1** Given that the payoffs in Exercise 3.2–1 are 2 or 4 dollars, how much should a person pay to draw a chip from the bowl to make this a "fair" game?

**3.2–2** To make the game in Exercise 3.2–3 fair, how much should a person pay to play?

## 3.3 The Mean, Variance, and Skewness

Let $X$ be a random variable of the discrete type with p.d.f. $f(x)$ and space $R$.

The mean of $X$ is defined by

$$\mu = E(X) = \sum_R x f(x).$$

The variance of $X$ is defined by

$$\sigma^2 = \mathrm{Var}(X) = E[(X-\mu)^2] = \sum_R (x-\mu)^2 f(x).$$

The standard deviation of $X$ is defined by

$$\sigma = \sqrt{\sigma^2} = \sqrt{\mathrm{Var}(X)}.$$

Skewness is defined by

$$A_3 = \frac{E[(X-\mu)^3]}{\{E[(X-\mu)^2]\}^{3/2}} = \frac{E[(X-\mu)^3]}{(\sigma^2)^{3/2}}.$$

When a distribution is symmetrical about its mean, the skewness is equal to zero. If the probability histogram has a "tail" to the right, $A_3 > 0$ and we say that the distribution is skewed positively or is skewed to the right. If the probability histogram has a "tail" to the left, $A_3 < 0$ and we say that the distribution is skewed negatively or is skewed to the left.

Let $x_1, x_2, \ldots, x_n$ be the observed values of a random sample of $n$ observations of $X$. The sample mean is defined by

$$\overline{x} = \frac{1}{n}\sum_{i=1}^{n} x_i.$$

The variance of the empirical distribution is defined by

$$\begin{aligned} v &= \frac{1}{n}\sum_{i=1}^{n}(x_i - \overline{x})^2 \\ &= \frac{1}{n}\sum_{i=1}^{n}x_i^2 - \overline{x}^2. \end{aligned}$$

The sample variance is defined by

$$\begin{aligned} s^2 &= \frac{1}{n-1}\sum_{i=1}^{n}(x_i - \overline{x})^2 \\ &= \frac{n\sum_{i=1}^{n}x_i^2 - \left(\sum_{i=1}^{n}x_i\right)^2}{n(n-1)}. \end{aligned}$$

## EXERCISES

**Purpose:** These exercises compare distributions that have different variances and distributions that have different skewness. The sample mean and sample variance are used to estimate a mean, $\mu$, and a variance, $\sigma^2$, respectively.

**3.3–1** Use `ProbHist` to depict a probability histogram for each of the following distributions. Find the values of $\mu$ and $\sigma^2$ for each distribution. Use *Maple*, if possible.

(a) $f(x) = (2- \mid x-4 \mid)/4, \; x = 3, 4, 5.$

(b) $f(x) = (4- \mid x-4 \mid)/16, \; x = 1, 2, 3, 4, 5, 6, 7.$

**3.3–2** For each of the following distributions find the values of $\mu$, $\sigma^2$, and $A_3$. Use *Maple*, if possible.

(a) $f(x) = \begin{cases} 2^{6-x}/64, & x = 1, 2, 3, 4, 5, 6, \\ 1/64, & x = 7. \end{cases}$

(b) $f(x) = \begin{cases} 1/64, & x = 1, \\ 2^{x-2}/64, & x = 2, 3, 4, 5, 6, 7. \end{cases}$

**3.3–3** Let $X$ equal the number of random numbers that must be added together so that their sum is greater than one.

(a) Simulate 10 observations of $X$. HINT: Let $U_1, U_2, \ldots$ be a sequence of random numbers from the interval $(0, 1)$. Then $X = \min\{k : \; u_1 + u_2 + \cdots + u_k > 1\}$. One could make 10 observations of $X$ by looking at `RNG(50);` or use the random numbers in Appendix C.

(b) Given that the p.d.f. of $X$ is $f(x) = (x-1)/x!$, $x = 2, 3, 4, \ldots$, use *Maple* to find $E(X)$ and $\mathrm{Var}(X)$.

(c) Simulate 200 observations of $X$. (We can note that $f(x)$ is a decreasing function and $\sum_{x=9}^{\infty} f(x) < 3 \times 10^{-5}$. Therefore this simulation can be done, in approximate form, through `DiscreteS((x-1)/x!,2 .. 8,200);`.) In order to show that the given p.d.f. is correct, the simulation can also be done using the following program and then comparing the empirical results with the claimed theoretical model.

```
for k from 1 to 200 do
XX[k] := 0;
s := 0;
while s < 1 do
s := s+rng(); XX[k] := XX[k]+1
od
od:
X := [seq(XX[k],k = 1 .. 200)];
```

(d) Is the sample mean, $\bar{x}$, close to $\mu = E(X)$?

(e) Is the sample variance, $s^2$, close to $\sigma^2 = \text{Var}(X)$?

(f) Use `ProbHist` and `Histogram` to plot a histogram of your 200 observations of $X$ with its p.d.f. superimposed.

**3.3–4** * Generate a sequence of random numbers from the interval $(0, 1)$: $u_1, u_2, u_3, \ldots$. If $u_1 \le u_2 \le \cdots \le u_k$ and $u_k > u_{k+1}$, let the random variable $X = k + 1$.

(a) Simulate 10 observations of $X$ using random numbers in Appendix C.

(b) Simulate 200 observations of $X$ and plot a histogram of these observations with the p.d.f. of $X$, $f(x) = (x - 1)/x!$, $x = 2, 3, 4, \ldots$, superimposed.

(c) Is the sample mean $\bar{x}$ close to $\mu = E(X) = e$?

**3.3–5** A Bingo card has 25 squares with numbers on 24 of them, the center being a free square. The numbers that are placed on the Bingo card are selected randomly and without replacement from 1 to 75, inclusive, for each card that is used. When a game called "cover-up" is played, balls numbered from 1 to 75, inclusive, are selected randomly and without replacement until a player **covers** each of the numbers on one of the cards being played.

(a) Suppose that you are playing this game with one card. Let $X$ equal the number of balls that must be drawn to "cover-up" all of the numbers on your card. Show that the p.d.f. of $X$ is

$$f(x) = \frac{\binom{24}{23}\binom{51}{x-24}}{\binom{75}{x-1}} \cdot \frac{1}{75 - (x - 1)}, \quad x = 24, 25, \ldots, 75.$$

(b) Find the mean, variance, and standard deviation of $X$.

(c) What value of $X$ is most likely?

(d) Suppose that $M$ cards are being played and let $Y$ equal the number of balls that must be selected to produce a winner. If the distribution function of $X$ is

$$F(x) = P(X \leq x), \quad x = 24, 25, \ldots, 75,$$

show that the distribution function of $Y$ is

$$G(y) = P(Y \leq y) = 1 - [1 - F(y)]^M, \quad y = 24, 25, \ldots, 75.$$

Thus the p.d.f. of $Y$ is

$$g(y) = P(Y = y) = [1 - F(y-1)]^M - [1 - F(y)]^M, \quad y = 24, 25, \ldots 75.$$

(e) Find the mean, variance, and standard deviation of $Y$.

(f) What value of $Y$ is most likely.

(g) Verify your theoretical derivations using simulation, comparing sample and theoretical characteristics.

**3.3–6** Consider each of the $n$ successive pairs of integers in a random permutation of $2n$ integers, the first and second, the third and fourth, the fifth and sixth, etc., as an interval. If necessary, switch the two numbers in a pair so the interval goes from small to large. Let $X$ equal the number of these intervals that intersect all others.

(a) Show, theoretically or empirically, that $P(X \geq 1) = 2/3$. That is, the probability is 2/3 that at least one interval intersects all others and this is true for all $n$.

(b) Show that, for $k < n$,

$$P(X \geq k) = \frac{2^k}{\binom{2k+1}{k}}.$$

(c) Show that $E(X) = \pi/2$. Hint: For a discrete random variable,

$$E(X) = \sum_{k=1}^{\infty} P(X \geq k).$$

**Remark:** This exercise is based on an article entitled "Random Intervals" by J. Justicz, E. R. Sheinerman, and P. M. Winkler in the December, 1990 issue of *The American Mathematical Monthly*, Vol. 94, No. 10, pp. 881–889.

## Questions and Comments

**3.3–1** In Exercise 3.3–1, what is the relation between $\sigma^2$ and the probability histogram?

**3.3–2** The mean and variance will be illustrated in future sections with many examples.

**3.3–3** In Exercise 3.3–3, is $\overline{x}$ close to $e$?

**3.3–4** In Exercise 3.3–3, **show** that the p.d.f. of $X$ is $f(x) = (x-1)/x!$, $x = 2, 3, 4, \ldots$.

# 3.4 The Hypergeometric Distribution

The p.d.f. of a hypergeometric random variable $X$ is

$$f(x) = \frac{\binom{n_1}{x}\binom{n_2}{r-x}}{\binom{n}{r}}$$

where $x \le r$, $x \le n_1$, $r - x \le n_2$, and $n_1 + n_2 = n$. The mean of $X$ is

$$\mu = E(X) = r\left(\frac{n_1}{n}\right)$$

and the variance of $X$ is

$$\sigma^2 = \text{Var}(X) = r\left(\frac{n_1}{n}\right)\left(\frac{n_2}{n}\right)\left(\frac{n-r}{n-1}\right).$$

## EXERCISES

**Purpose:** Hypergeometric probabilities are illustrated empirically.

**3.4–1** Draw an unordered sample of size five from a deck of playing cards, sampling without replacement.

(a) What is the probability of drawing exactly two hearts?

(b) Illustrate this empirically by taking 20 samples of size five from a deck of cards. Return the five cards to the deck and shuffle the deck between draws. Record your results in Table 3.4–1.

| | Number of Occurrences | | Relative Frequency | | |
|---|---|---|---|---|---|
| | For Me | For Class | For Me | For Class | Probability |
| Exactly 2 Hearts | | | | | |
| Other Hands | | | | | |
| Total | | | | | |

**Table 3.4–1**

**3.4–2** Simulate 200 observations of drawing five card hands from a deck of cards and counting the number of hearts in each hand. You may use

```
X := HypergeometricS(13,39,5,200);
```

or the following:

```
with(combinat,randcomb);
for k from 1 to 200 do
S := randcomb([1 $ 13,0 $ 39],5);
XX[k] := sum(S[i],i = 1 .. 5)
od:
X := [seq(XX[i],i = 1 .. 200)];
```

Use `Freq(X,0..5);` to find the frequencies of the possible outcomes. Also make a histogram of the data with the hypergeometric probability histogram superimposed, `y := binomial(13,x)*binomial(39,5-x)/binomial(52,5);`.

**3.4–3** Simulate 50 draws of five card hands. The *Maple* supplement has the data structure

```
Cards := [C1,...,CJ,CQ,CK,D1,...,DK,H1,...,HK,S1,...,SK]
```

available for your use. To simulate a 5 card hand, simply use `randcomb(Cards,5);`

(a) For your set of 50 hands, fill in a table like Table 3.4–1.

(b) For your set of 50 hands, fill in Table 3.4–2.

| | Number of Occurrences | | Relative Frequency | | |
|---|---|---|---|---|---|
| | For Me | For Class | For Me | For Class | Probability |
| Three of a kind, + 2 others | | | | | |
| Two pairs + 1 other | | | | | |
| One pair + 3 others | | | | | |
| Totals | | | | | |

**Table 3.4–2**

**3.4–4** Select six integers between 1 and 47, inclusive. Now make a random selection of six integers between 1 and 47, inclusive, using *Maple*. Let $A_i$, $i = 0, 1, \ldots, 6$, be the events that for your six integers and the computer's six integers, exactly $i$ integers match.

(a) Find $P(A_i)$, $i = 0, 1, \ldots, 6$.

(b*) Simulate this problem 100 times, comparing the proportions of times that $A_i$ is observed with $P(A_i)$, $i = 0, 1, 2, \ldots, 6$, respectively. Do this by plotting the probability histogram and the empirical histogram on the same graph. Note that, without loss of generality, we may select as our integers 1, 2, 3, 4, 5, 6. Then we can use the following simulation:

```
with(combinat,randcomb);
for i from 1 to 100 do
XX[i] := 0;
L := randcomb(47,6);
for j from 1 to 6 do
if L[j] < 7 then XX[i] := XX[i]+1 fi
od
od:
X := [seq(XX[i],i = 1 .. 100)];
```

(c) Compare the distribution mean and the sample mean.

(d) Compare the distribution variance and the sample variance.

**3.4–5** Suppose that an urn contains $2n$ balls labeled $S$ and $2n$ balls labeled $J$. Randomly select $2n$ balls from the urn, one at a time, without replacement. Let $X$ equal the absolute value of the difference of the number of $S$ balls and the number of $J$ balls that are selected.

(a) Define the p.d.f., $f(x)$, of $X$ (either for a general $n$ or for $n = 10$).

(b) Show that the sum of all the probabilities is equal to 1.

(c) Find the mean and the variance of $X$.

(d*) Simulate 200 observations of $X$ for $n = 10$ and compare theoretical and empirical characteristics, such as means, variances, and probabilities.

```
g := x -> binomial(2*n,n+1/2*x)*
binomial(2*n,n-1/2*x)/binomial(4*n,2*n);
f := proc(x)
if x = 0 then g(0)
elif x mod 2 = 0 then 2*g(x)
else 0
fi
end;
randomize();
n := 10;
for k from 1 to 200 do
```

```
S := 2*n;
J := 2*n;
for j to 2*n do
vote := floor(S/(S+J)+rng());
if vote = 1 then S := S-1
else J := J-1
fi
od;
XX[k] := abs(S-J)
od:
X := [seq(XX[k],k = 1 .. 200)];
```

**Remark**: This exercise is based on Problem 10248 in *The American Mathematical Monthly*, October, 1992, page 781. The problem is:

> Candidates Smith and Jones are the only two contestants in an election that will be deadlocked when all the votes are counted—each will receive $2n$ of the $4n$ votes cast. The ballot count is carried out with successive random selections from a single container. After exactly $2n$ votes are tallied, Smith has $S$ votes and Jones has $J$ votes. What is the expected value of $|S - J|$?

## Questions and Comments

**3.4–1** Exercise 3.4–4 is based on a state lottery game called LOTTO47. What is your recommendation about placing a bet in LOTTO47?

**3.4–2** There are variations of LOTTO47, e.g., selecting numbers from a larger set of numbers. In 1993 Michigan started selecting six integers out of 1 to 49, inclusive. For a charge of $1 (which is never returned), the prize for matching all six integers is at least $2,000,000. The prize is $2,500 for matching five and $100 for matching four of the State's six selections. What is the expected payoff to the player?

# 3.5 Bernoulli Trials and The Binomial Distribution

A Bernoulli experiment is a random experiment, the outcome of which can be classified in but one of two mutually exclusive and exhaustive ways, say success and failure. A sequence of Bernoulli trials occurs when a Bernoulli experiment is performed several independent times and the probability of success, $p$, remains the same from trial to trial. We let $q = 1 - p$, the probability of failure.

Let $X$ be a random variable associated with a Bernoulli experiment. The p.d.f. of $X$ is

$$f(x) = p^x(1-p)^{1-x} = p^x q^{1-x}, \; x = 0, 1.$$

The mean and variance of $X$ are $\mu = p$ and $\sigma^2 = p(1-p) = pq$.

Probabilities for Bernoulli distributions can be found by using `BernoulliPDF(p,x);`. If you specify only `p`, as in `BernoulliPDF(0.3,x);`, then the Bernoulli p.d.f. is returned as a function of `x`; if you give a specific value to `x`, the probability for that value of `x` is returned. In a similar way, `BernoulliCDF(p,x);` returns the distribution function of a Bernoulli random variable.

In a sequence of $n$ Bernoulli trials, let $Y_i$ denote the Bernoulli random variable associated with the $i$th trial. If we let $X$ equal the number of successes in $n$ Bernoulli trials, then

$$X = \sum_{i=1}^{n} Y_i$$

and $X$ has a binomial distribution. We say that $X$ is $b(n,p)$. The p.d.f., mean, and variance of $X$ are

$$f(x) = \frac{n!}{x!(n-x)!} p^x (1-p)^{n-x}, x = 0, 1, 2, \ldots, n,$$

$\mu = np$, and $\sigma^2 = np(1-p) = npq$.

Probabilities for binomial distributions can be found by using `BinomialPDF(n,p,x);`. If you specify only `n` and `p`, as in `BinomialPDF(20,0.3,x);`, then the binomial p.d.f. is returned as a function of `x`; if you give a specific value to `x`, the probability for that value of `x` is returned.

**Remark.** All of the arguments for `BernoulliPDF`, `BernoulliCDF`, `BinomialPDF`, and `BinomialCDF` can be either variable names or numeric values.

## EXERCISES

**Purpose:** The exercises illustrate the relation between Bernoulli trials and the binomial distribution. They also illustrate the relation between relative frequency and probability of events for the binomial distribution.

**3.5–1** An urn contains 6 red and 4 white balls. One ball is to be drawn from the urn with replacement with red denoting success and white denoting failure.

(a) Use `BernoulliS(0.6,100);` to simulate 100 trials.

(b) Calculate the relative frequency of success. Is this number close to 6/10?

**3.5–2** Almost all statistics books have tables for binomial probabilities and cumulative binomial probabilities. Use `BinomialPDF(10,0.3,4);` to obtain $P(X = 4)$ when $X$ is $b(10, 0.3)$ and check if the value of $P(X = 4)$ given in your book is correct. Use `BinomialCDF(10,0.3,4);` to do the same for $P(X \leq 4)$.

**3.5–3** The Bernoulli p.d.f. makes sense for $0 \leq p \leq 1$ and $x = 0$ and 1. However, if interpreted as a function of $p$ and $x$, both on the $[0,1]$ interval, it can be plotted as a surface. Use `y := BernoulliPDF(p,x);` and `plot3d(y,p = 0 .. 1,x = 0 .. 1,` `axes = BOXED);` to obtain a graph of the surface described by `y`. For what value(s) of $p$ is $P(X = 0) = P(X = 1)$? Can you verify your answer algebraically?

**3.5–4** As was the case for the Bernoulli p.d.f. in Exercise 3.5–3, the binomial p.d.f. can be considered as a function of $n$, $p$ and $x$ (again, extending the original specifications that mandate that $x$ be an integer between 0 and $n$). For $n = 20$, obtain a 3-dimensional plot ($0 \le p \le 1$ and $0 \le x \le 20$) of the binomial p.d.f. Given $p$, for what value of $x$ does the binomial probability seem maximal?

**3.5–5** If you are using *Maple* in a windows environment (on a Macintosh computer, Microsoft Windows in DOS or X-Windows), you can look at the binomial p.d.f. for a fixed $n$, say $n = 10$, and varying $p$. The following will animate the probability Histogram of the binomial distribution.

```
for i from 1 to 9 do
f := BinomialPDF(10, i/10, x):
H.i := ProbHist(f, 0..10):
od:
display([seq(H.i,i = 1 .. 9)],insequence = true);
```

How is the shape of the binomial probability histogram affected by $p$?

**3.5–6** Modify what was done in Exercise 3.5–5 to animate the binomial probability histogram for a fixed $p$, say $p = 1/4$, and $n$ between 2 and 20. How is the shape of the binomial probability histogram affected by $n$?

**3.5–7** Use the fact that if $X$ is $b(n, p)$, then $X$ is equal to the sum of $n$ independent and identically distributed Bernoulli random variables with $p$ equal to the probability of success on each trial.

(a) Simulate a random sample of size 200 from a binomial distribution $b(7, 0.6)$. Each observation of $X$ is the sum of 7 Bernoulli trials. Count the number of times that each outcome was observed using `Freq(X,0 .. 7);`. For this simulation, use

```
randomize();
p := .6 ;
n := 7;
for k from 1 to 200 do
XX[k] := 0;
for j from 1 to n do
XX[k] := XX[k]+floor(rng()+p)
od
od:
X := [seq(XX[k],k = 1 .. 200)];
```

or

```
B := [seq(BernoulliS(.6 ,7),i = 1 .. 200)]:
X := [seq(sum(B[i][j],j = 1 .. 7),i = 1 .. 200)];
```

(b) Let $X$ have a binomial distribution $b(7, 0.6)$. Compare the theoretical and empirical probabilities graphically using

```
y := BinomialPDF(7,.6,x);
p1 := ProbHist(y,0 .. 7):
p2 := Histogram(X,-.5  .. 7.5,8):
plot({p1,p2});
```

(c) Compare the theoretical and empirical distribution functions graphically using `BinomialCDF`, `PlotEmpCDF`, and `PlotDiscCDF`.

(d) Compare the sample and distribution means, the sample and distribution variances.

**Remark:** In the future, you may use `X := BinomialS(7,0.6,200);`.

**3.5–8** Use *Maple* to verify that for a given $n$ and $p$, the binomial probabilities sum to 1. In the case of a particular $n$ and $p$, say $n = 25$ and $p = 0.3$, the binomial probabilities can be expressed by `y := BinomialPDF(25,0.3,x);` and the sum of the probabilities can be obtained through `s := sum(y,x = 0 .. 25);`. If necessary, the expression for `s` can be simplified through `simplify(s);`. Take advantage of the fact that `BinomialPDF` accepts symbolic as well as numeric arguments, to generalize this to a "proof" that for any $X$ that is $b(n, p)$, $\sum_{x=0}^{n} P(X = x) = 1$.

**3.5–9** Through a process similar to that used in the last exercise, show that if $X$ is $b(n, p)$ then $\mu = np$ and $\sigma^2 = np(1 - p)$.

**3.5–10** Let $X$ denote the number of heads occurring on a single toss of 6 unbiased coins.

(a) Simulate 100 observations of $X$.

(b) Graphically compare the theoretical and empirical histograms.

(c) Graphically compare the theoretical and empirical distribution functions.

**3.5–11** Let $X$ denote the number of 1's and 2's occurring on a single roll of 8 unbiased dice.

(a) Interpret $X$ as a binomial random variable and simulate 200 observations of $X$.

(b) Compare the probabilities and relative frequencies of the nine possible outcomes.

(c) Compare cumulative relative frequencies with the distribution function.

## Questions and Comments

**3.5–1** Let $X$ be $b(3,1/4)$. Then $P(X = 0) = 27/64$, $P(X = 1) = 27/64$, $P(X = 2) = 9/64$, $P(X = 3) = 1/64$. Describe how one random number can be used to simulate an observation of $X$. This is essentially what is used in the following:

```
L := [0,27/64,1,27/64,2,9/64,3,1/64];
X := DiscreteS(L,200);
```

# 3.6 Geometric and Negative Binomial Distributions

Let a random experiment be the observation of a sequence of Bernoulli trials with probability of success $p$ on each trial. Let $X$ denote the trial number on which the first success occurs. The p.d.f. of $X$ is

$$f(x) = p(1-p)^{x-1}, \ x = 1, 2, \dots .$$

We say that $X$ has a geometric distribution. The mean and variance of $X$ are given by $\mu = 1/p$ and $\sigma^2 = (1-p)/p^2 = q/p^2$.

Suppose now that the sequence of Bernoulli trials is observed until exactly $r$ successes occur, $r$ a fixed positive integer. Let $X$ denote the number of trials needed to observe exactly $r$ successes. The p.d.f. of $X$ is

$$g(x) = \binom{x-1}{r-1}(1-p)^{x-r}p^r, \ x = r, r+1, \dots .$$

We say that $X$ has a negative binomial distribution. The mean and variance of $X$ are $\mu = r/p$ and $\sigma^2 = r(1-p)/p^2 = rq/p^2$.

Probabilities for the geometric and negative binomial distributions can be found by using `GeometricPDF(p,x);` and `NegBinomialPDF(p,r,x);`, respectively. Cumulative probabilities are accessed through `GeometricCDF(p,x);` and `NegBinomialCDF(p,r,x);`, respectively. All arguments to any of these procedures can be either variable names or numeric values.

## EXERCISES

**Purpose:** The exercises illustrate the relation between Bernoulli trials and the geometric and negative binomial distributions.

**3.6–1** An experiment consists of rolling a 4-sided die until a one is observed. Let $X$ equal the number of trials required.

(a) Fill in the probabilities in Table 3.6–1.

(b) Perform this experiment 20 times, recording your frequencies in Table 3.6–1.

(c) Use `Die(4,60);` to find 10 observations of $X$.

(d*) Simulate this experiment 200 times and obtain results like those asked for in Table 3.6–1. That is, observe a sequence of Bernoulli trials for which the probability of success is $p = 0.25$ on each of the independent trials. Count the number of trials that are needed to observe a success. Repeat this process 200 times.

```
randomize();
p := .25;
for k from 1 to 200 do
XX[k] := 0;
success := 0;
while success < 1 do
XX[k] := XX[k]+1;
success := floor(rng()+p)
od
od:
X := [seq(XX[k],k = 1 .. 200)];
```

(e) Use `Y := GeometricS(0.25,200);` to simulate 200 trials of this experiment. Obtain results like those asked for in Table 3.6–1.

| | Frequencies | | Relative Frequencies | | |
|---|---|---|---|---|---|
| $X$ | For Me | For Class | For Me | For Class | Probability |
| 1 | | | | | |
| 2 | | | | | |
| 3 | | | | | |
| 4 | | | | | |
| $\geq 5$ | | | | | |

**Table 3.6–1**

(f) For each set of data that you generated, on the same graph plot the relative frequency histogram of the data and the probability histogram; on the same graph plot the empirical distribution function and the theoretical distribution function.

(g) Construct a box-and-whisker display of your data and interpret it.

**3.6–2** Suppose that there is a payoff associated with the experiment described in Exercise 3.6–1. In particular let the payoff be equal to $x$ dollars. That is, if $x$ trials are needed to observe the first one, the payoff is $x$ dollars.

(a) Define the expected value of payment in terms of a sum and give the value of this sum. (You may use `sum(.25*x*.75^(x-1),x = 1 .. infinity);`.) How is this sum related to $\mu = E(X)$?

(b) Find the average of the observed values in Exercises 3.6–1(b), (d), and (e). Are these averages close to your answer in Exercise 3.6–2(a)?

(c) Compare the sample variance(s) with the theoretical variance for the data that you generated in Exercise 3.6–1.

**3.6–3** Let $X$ equal the number of times that a coin must be flipped to observe heads. That is, $X$ is a geometric random variable with $p = 0.5$. Thus $\mu = E(X) = 2$. Suppose now that the payoff is equal to $2^X$.

(a) Show that $E(2^X) = \infty$.

(b) Illustrate this last result empirically. In particular, simulate 500 or 1000 repetitions of this experiment. Let `X` be a list that holds the outcomes. Store the values of $2^{X[k]}$ as a list, say `P` (for Payoff), and graph a running average of the payoffs. Interpret your output.

(c) Draw a box-and-whisker display of the payoffs. Again, interpret your output.

**3.6–4** Interpret the p.d.f. of the Geometric distribution as a function of $p$ and $x$ and get a 3-dimensional plot for $0 \leq p \leq 1$ and $1 \leq x \leq 10$. Regardless of the value of $p$, what happens to the p.d.f. when $x$ gets large?

**3.6–5** Use *Maple* to verify that the probabilities of a geometric random variable sum to 1.

**3.6–6** Use *Maple* to show that the mean and variance of a geometric random variable are $1/p$ and $(1-p)/p^2$, respectively.

**3.6–7** If you are using *Maple* in a windows environment (on a Macintosh computer, Microsoft Windows in DOS or X-Windows), you can animate the probability histogram of the geometric p.d.f. (see Exercise 3.5–5). Animate the geometric probability histogram and describe how it is affected by $p$.

**3.6–8** Use *Maple* to verify that the probabilities of a negative binomial random variable sum to 1.

**3.6–9** Use *Maple* to show that the mean and variance of a negative binomial random variable are $r/p$ and $r(1-p)/p^2$, respectively.

**3.6–10** An experiment consists of flipping $n = 8$ fair coins until heads has been observed on each coin. That is, the 8 coins are flipped and all heads are removed. The coins that had landed tails are flipped again and all heads from this group are removed. This procedure is continued until each coin has come up heads. Let the random variable $X$ equal the number of tosses required to complete this process. If $Y_i$ is the number of trials needed for coin $i$ to land heads, then $Y_i$ has a geometric distribution with $p = 1/2, i = 1, 2, \ldots, 8$, and

$$X = \max\{Y_1, Y_2, \ldots, Y_8\}.$$

Furthermore, if $x$ is a positive integer, then the distribution function of $X$ is given by

$$\begin{aligned} F(x) &= P(X \leq x) \\ &= P(\max\{Y_1, Y_2, \ldots, Y_8\} \leq x) \\ &= [P(Y_1 \leq x)]^8 \\ &= [1 - (1/2)^x]^8. \end{aligned}$$

(a) Show that the p.d.f. of $X$ is defined by

$$f(x) = [1 - (1/2)^x]^8 - [1 - (1/2)^{x-1}]^8, \; x = 1, 2, 3, \ldots.$$

(b) Find the value of $E(X)$.

(c) Find the value of $\text{Var}(X)$.

(d) Simulate 100 observations of $X$. For example, you could use the *Maple* command `X := [seq(Max(GeometricS(0.5,8)),i = 1 .. 100)];`. Compare your answers to parts (b) and (c) with the sample mean, $\overline{x}$, and the sample variance, $s^2$, respectively.

(e) Plot a relative frequency histogram of your data with the p.d.f. defined in part (a) superimposed.

(f) Work this problem using either more coins or fewer coins. For example, use 4 or 16 coins. Especially note the effect of the number of coins on $E(X)$ when the number of coins is doubled or halved. Do your answers make sense intuitively?

**Remark.** This problem was motivated by the article "Tossing Coins Until All are Heads," *Mathematics Magazine*, May, 1978, pages 184–186, by John Kenny.

**3.6–11** Modify Exercise 3.6–10 by rolling 8 dice until a one has been observed on each die. Let $X$ equal the number of rolls required.

**3.6–12** Notice that the p.d.f of Exercise 3.6–11 is identical to that of Exercise 3.6–10, except that the 1/2 is replaced by 1/6. Consider the more general situation where an arbitrary $p$ with $0 \leq p \leq 1$ is used. From a 3-dimensional plot of the resulting p.d.f. (perhaps several plots) with $p$ in $[0, 1]$ and $x \geq 1$, determine how $E(X)$ changes with increasing $p$. How does $\text{Var}(X)$ change with increasing $p$?

**3.6–13** Modify Exercise 3.6–11 by rolling 12 dice until a one has been observed on each die. Let $X$ equal the number of rolls required.

**3.6–14** Suppose we have $n$ identical coins for each of which heads occurs with probability $p$. Suppose we first toss all the coins, then toss those which show tails after the first toss, then toss those which show tails after the second toss, and so on until all the coins show heads. Let $X$ be the number of coins involved in the last toss. Either empirically or theoretically or both:

(a) Show that

$$\begin{aligned} P(X=x) &= \sum_{k=1}^{\infty}\binom{n}{x}(q^{k-1}p)^x(1-q^{k-1})^{n-x} \\ &= p^x\binom{n}{x}\sum_{j=0}^{n-x}\binom{n-x}{j}\frac{(-1)^j}{1-q^{j+x}}, \quad x=1,2,\ldots,n-1. \\ P(X=1) &= np\sum_{j=0}^{n-1}\binom{n-1}{j}\frac{(-1)^j}{1-q^{j+1}}. \end{aligned}$$

(b) Show that $E(X) = P(X=1)/q$.

(c) Let $p_n = P(X=1)$. Analyze the behavior of $p_n$ as $n \to \infty$. In particular, show that $p_n$ oscillates around the pseudo-limit $-p/\log q$.

**Remark.** This exercise is based on Problem 3436 in *The American Mathematical Monthly,* Vol. 98, No. 4, p. 366, by Lennart Räde (April, 1991). Its solution appears in *The American Mathematical Monthly,* Vol. 101, No. 1 (January, 1994), pp. 78–80.

**3.6–15** For Exercise 3.6–10, let $X$ equal the number of dice that are rolled on the last roll. Answer questions similar to those in Exercise 3.6–14.

**3.6–16** An urn contains 3 red and 7 blue marbles. Marbles are drawn at random from the urn, one at a time, with replacement. Let red denote success and blue failure. Let $X$ denote the number of trials needed to observe the first success.

(a) Simulate 100 observations of $X$.

(b) Find the average of these 100 observations.

(c) If there is a payoff of $x$ dollars if $x$ trials are needed, find the expected value of payment. That is, find $E(X)$.

(d) Are your answers in parts (b) and (c) approximately equal?

**3.6–17** An experiment consists of flipping a coin until the third head is observed. Let $x$ equal the number of flips required.

(a) In Table 3.6–2 fill in the probabilities.

(b) Perform this experiment 20 times, recording your frequencies and relative frequencies in Table 3.6–2.

(c) Use `NegBinomialS(r,p,200);` or modify the program in Exercise 3.6-1 to simulate 200 trials of this experiment. Obtain results like those asked for in Table 3.6–2. (See part (e).)

| | Frequencies | | Relative Frequencies | | |
|---|---|---|---|---|---|
| $X$ | For Me | For Class | For Me | For Class | Probability |
| 3 | | | | | |
| 4 | | | | | |
| 5 | | | | | |
| 6 | | | | | |
| 7 | | | | | |
| 8 | | | | | |
| 9 | | | | | |

**Table 3.6–2**

(d) If there is a payoff of $x$ dollars if $x$ flips are needed, find the expected value of payment. That is, find $\mu = E(X)$.

(e) Find the average of the 200 outcomes in part (c). Is this average close to $\mu = E(X)$?

(f) Graph the histogram of your observations along with the p.d.f. for the negative binomial distribution.

(g) Graph the empirical and theoretical distribution functions.

**3.6–18** Seven different prizes, called freakies, were randomly put into boxes of cereal, one per box. A family decided to buy this cereal until they had obtained at least one of each of the seven different freakies. Given that $i-1$ freakies had been obtained, let $X_i$ denote the number of additional boxes of cereal that must be purchased to obtain a different freakie, $i = 1, 2, \ldots, 7$. Note that $X_1 = 1$.

(a) Show that $X_i$ has a geometric distribution with probability of success $p_i = (8-i)/7,\ i = 1, 2, \ldots, 7$.

(b) Find $E(X_i),\ i = 1, 2, \ldots, 7$.

(c) Let $Y = X_1 + X_2 + \cdots + X_7$. Find $E(Y)$, the expected number of boxes of cereal that must be purchased to obtain each freakie at least once.

(d) Simulate 100 observations of $Y$. Compare the average of these observations of $Y$ with $E(Y)$.

(e) Draw a histogram of your observations to estimate the p.d.f. of $Y$.

**Remark:** Note that each $X_i$ has a geometric distribution with probability of success $p_i = (8-i)/7$, $i = 1, 2, \ldots, 7$. In Section 5.2 methods are given for finding the p.d.f. of $Y$, either using the convolution formula successively or by using moment-generating functions.

**3.6–19** Consider the experiment of flipping a fair coin until the same face is observed on successive flips. Let $X$ equal the number of flips that are required.

(a) Define the p.d.f. of $X$.

(b) Find the mean and variance of $X$.

(c) Simulate this experiment (perhaps first physically by actually flipping coins) and compare your empirical results with the answers in parts (a) and (b). (See Exercise 1.1–6.)

**3.6–20** Consider the experiment of flipping a fair coin until heads is observed on successive flips. Let $X$ equal the number of flips that are required.

(a) Define the p.d.f. of $X$.

(b) Find the mean and variance of $X$.

(c) Simulate this experiment (perhaps first physically by actually flipping coins) and compare your empirical results with the answers in parts (a) and (b). Hint: See Exercise 1.1–7. Also, to define the p.d.f. you could use `with(combinat, fibonacci): y := fibonacci(x - 1)/2^x;`. Why?

## Questions and Comments

**3.6–1** Describe some experiments for which the geometric and/or negative binomial distributions would be appropriate.

# 3.7 The Poisson Distribution

The p.d.f. of a Poisson random variable, $X$, is given by

$$f(x) = \frac{\lambda^x e^{-\lambda}}{x!}, \quad x = 0, 1, 2, \ldots.$$

The parameter $\lambda$ is equal to both the mean and the variance of $X$. That is $\mu = \sigma^2 = \lambda$. In general $\lambda$ is equal to the average number of changes (events) observed per unit. If $\lambda$ is the average number of changes observed per unit, then $\lambda s$ is the average number of changes observed in s units.

The Poisson probabilities can be used to approximate binomial probabilities. If $X$ has a binomial distribution $b(n, p)$, then

$$\binom{n}{x} p^x (1-p)^{n-x} \approx \frac{(np)^x e^{-np}}{x!}.$$

This approximation is good if $p$ is "small" and $n$ is "large." A rule of thumb says that the approximation is quite accurate if $n \geq 20$ and $p \leq 0.05$; it is very good if $n \geq 100$ and $np \leq 10$.

The procedures `PoissonPDF(lambda,x);` and `PoissonCDF(lambda,x);` give the probabilities and cumulative probabilities of a Poisson random variable with $\lambda =$`lambda`. The arguments `lambda` and `x` can be numeric or symbolic.

## EXERCISES

**Purpose:** The exercises illustrate the shape of the p.d.f. of a Poisson random variable. A data set is given for an experiment for which the Poisson distribution is appropriate. Observations of a Poisson random variable are simulated with `PoissonS` and with observations of a binomial random variable. The rule of thumb for approximating binomial probabilities with Poisson probabilities is investigated.

**3.7–1** Let $X$ be a Poisson random variable with $\lambda = 1.3$.

(a) Use `PoissonPDF` to find $P(X = 3)$. Is $P(X = 3)$ given in the table in your book correct?

(b) Use `PoissonCDF` to find $P(X \leq 3)$. Is $P(X \leq 3)$ given in the table in your book correct?

**3.7–2** Use `ProbHist` to depict the p.d.f. as a probability histogram for the Poisson distribution for (a) $\lambda = 0.7$, (b) $\lambda = 1.3$, (c) $\lambda = 2.7$, and (d) $\lambda = 5.59$.

**3.7–3** Consider the p.d.f. of a Poisson random variable, $X$, as a function of $\lambda$ and $x$ and plot it for $0 \leq x \leq 20$ and $0 \leq \lambda \leq 10$. It should be obvious that Var($X$) increases as $\lambda$ increases. How does the skewness of $X$ change with increasing $\lambda$? How does $M = \max_{x \geq 0}\{P(X = 0),\ P(X = 1),\ P(X = 2), \ldots\}$ change with increasing $\lambda$?.

**3.7–4** Use *Maple* to show that the probabilities of a Poisson random variable sum to 1.

**3.7–5** Use *Maple* to verify that the mean and variance of a Poisson random variable satisfy $\mu = \sigma^2 = \lambda$.

**3.7–6** If $X$ is the number of alpha particles emitted by a radioactive substance that enter a region during a fixed period of time, then $X$ is approximately Poisson. Table 3.7–1 lists 100 observations of the number of emissions of Barium-133 in 1/10 of a second and counted by a Geiger counter in a fixed position (these are stored in the file `STAT.DAT` under the name `D376` and can be accessed by `read 'STAT.DAT';`).

| 7 | 5 | 7 | 7 | 4 | 6 | 5 | 6 | 6 | 2 |
|---|---|---|---|---|---|---|---|---|---|
| 4 | 5 | 4 | 3 | 8 | 5 | 6 | 3 | 4 | 2 |
| 3 | 3 | 4 | 2 | 9 | 12 | 6 | 8 | 5 | 8 |
| 6 | 3 | 11 | 8 | 3 | 6 | 3 | 6 | 4 | 5 |
| 4 | 4 | 9 | 6 | 10 | 9 | 5 | 9 | 5 | 4 |
| 4 | 4 | 6 | 7 | 7 | 8 | 6 | 9 | 4 | 6 |
| 5 | 3 | 8 | 4 | 7 | 4 | 7 | 8 | 5 | 8 |
| 3 | 3 | 4 | 2 | 9 | 1 | 6 | 8 | 5 | 8 |
| 5 | 7 | 5 | 9 | 3 | 4 | 8 | 5 | 3 | 7 |
| 3 | 6 | 4 | 8 | 10 | 5 | 6 | 5 | 4 | 4 |

**Table 3.7–1**

Denote the observations by $x_1, x_2, \ldots, x_{100}$ and find the sample mean $\overline{x}$ and the sample variance $s^2$. Compare the probabilities and relative frequencies of the various outcomes by plotting a relative frequency histogram with a probability histogram superimposed.

**3.7–7** (a) Use `PoissonS(5.6, 100);` to obtain 100 observations of a Poisson random variable having a mean $\lambda = 5.6$.

(b) Depict a relative frequency histogram with the Poisson probability histogram superimposed.

(c) Are both $\overline{x}$ and $s^2$ approximately equal to 5.6?

(d) Plot the empirical distribution function with the superimposed theoretical distribution function.

**3.7–8** (a) For the Poisson distribution, the mean and variance are equal. For the binomial distribution, does $np \approx np(1-p)$ for those values of $n$ and $p$ for which the rule of thumb claims that Poisson probabilities can be used to approximate binomial probabilities? Answer this question for several values of $n$ and $p$.

(b) For $n = 20$ and $p = 0.05$, compare the binomial probabilities with the Poisson approximations, that is, compare

$$\binom{20}{x}(0.05)^x(0.95)^{20-x} \quad \text{with} \quad \frac{1^x e^{-1}}{x!}$$

for $x = 0, 1, \ldots, 20$. Note that the calculations mandated by the above expressions can be done via `BinomialPDF` and `PoissonPDF`.

(c) Repeat part (b) for other values of $n$ and $p$.

(d) For a fixed $n$, say $n = 20$, $b(n, p)$ can be approximated by the Poisson distribution with $\lambda = np = 20p$. Study the quality of this approximation by plotting (use `plot3d` here) the difference of the two p.d.f.'s for appropriate ranges of $p$ and $x$. Does the rule of thumb for approximating a binomial by a Poisson hold?

**3.7–9** (a) We shall use the observations of a binomial random variable $X$, $b(n,p)$, to simulate an observation of a Poisson random variable with a mean $\lambda = 5.6$. In particular let $X$ be $b(100,0.056)$ and generate 100 observations of $X$ to simulate 100 observations of a Poisson random variable with $\lambda = 5.6$.

(b) Plot the relative frequency histogram of the sample obtained in part (a) with the probability histogram of the appropriate Poisson distribution. (See Exercise 3.7–7.)

(c) Are both $\overline{x}$ and $s^2$ approximately equal to 5.6?

**3.7–10** A Geiger counter was set up in the physics laboratory to record the number of alpha particle emissions of carbon 14 in 0.5 seconds. Table 3.7–2 lists 150 observations (available as `D3710` in `STAT.DAT`).

| | | | | | | | | | | | | | | |
|---|---|---|---|---|---|---|---|---|---|---|---|---|---|---|
| 9 | 8 | 5 | 6 | 8 | 9 | 4 | 7 | 7 | 6 | 5 | 11 | 11 | 13 | 11 |
| 7 | 10 | 8 | 13 | 13 | 6 | 9 | 7 | 6 | 10 | 12 | 9 | 8 | 9 | 12 |
| 8 | 10 | 9 | 9 | 10 | 17 | 9 | 8 | 10 | 7 | 3 | 6 | 7 | 15 | 8 |
| 9 | 12 | 10 | 13 | 5 | 4 | 6 | 6 | 12 | 11 | 11 | 10 | 5 | 10 | 7 |
| 9 | 6 | 9 | 11 | 9 | 6 | 4 | 9 | 11 | 8 | 10 | 11 | 7 | 4 | 5 |
| 8 | 6 | 5 | 8 | 7 | 6 | 3 | 4 | 6 | 4 | 12 | 7 | 14 | 5 | 9 |
| 7 | 10 | 9 | 6 | 6 | 12 | 10 | 7 | 12 | 13 | 7 | 8 | 8 | 9 | 11 |
| 4 | 14 | 6 | 7 | 11 | 8 | 6 | 3 | 8 | 8 | 8 | 10 | 11 | 8 | 11 |
| 8 | 12 | 6 | 17 | 10 | 9 | 13 | 8 | 6 | 3 | 9 | 6 | 8 | 6 | 13 |
| 8 | 6 | 8 | 13 | 9 | 13 | 8 | 8 | 13 | 7 | 4 | 13 | 7 | 13 | 7 |

**Table 3.7–2**

(a) Let $x_1, x_2, \ldots, x_{150}$ denote these observations. Calculate the value of the sample mean, $\overline{x}$.

(b) To compare the probabilities for a Poisson random variable with mean $\lambda = \overline{x}$, plot the relative frequency histogram of $x_1, x_2, \ldots, x_{150}$ superimposed on the appropriate Poisson probability histogram.

(c) Are $\overline{x}$ and $s^2$ approximately equal?

**3.7–11** If you are using *Maple* in a windows environment (on a Macintosh computer, Microsoft Windows in DOS or X-Windows), you can animate the probability histogram of the Poisson distribution. (See Exercises 3.5–5 and 3.6–7.) Animate the geometric probability histogram and describe how it is affected by $\lambda$.

**3.7–12** Given a random permutation of the first $n$ positive integers, let $X$ equal the number of integers in their natural position. When $n$ is "sufficiently large" the random variable $X$ has an approximate Poisson distribution with $\lambda = 1$. Use simulation to determine how large $n$ should be for this to be a good fit. (See Exercise 2.1–6(b).)

**3.7–13** Roll an $n$-sided die $n$ times and let $X$ equal the number of rolls on which face $k$ is rolled on the $k$th roll, $k = 1, 2, \ldots, n$. Does $X$ have an approximate Poisson distribution when $n$ is sufficiently large?

## Questions and Comments

**3.7–1** Compare the data for Exercises 3.7–6, 3.7–7, and 3.7–9. Are they similar?

# 3.8 Moment-Generating Functions

Let $X$ be a random variable of the discrete type with p.d.f. $f(x)$ and space $R$.

If there is a positive number $h$ such that

$$M(t) = E(e^{tX}) = \sum_x e^{tx} f(x)$$

exists for $-h < t < h$, then $M(t) = E(e^{tX})$ is called the moment-generating function of $X$. It can be used to find

$$\mu = E(X) = M'(0),$$
$$\sigma^2 = E(X^2) - [E(X)]^2 = M''(0) - [M'(0)]^2.$$

When the moment-generating function exists, derivatives of all orders exist at $t = 0$. It is then possible to represent $M(t)$ as a Maclaurin's series, namely

$$M(t) = M(0) + M'(0)\frac{t}{1!} + M''(0)\frac{t^2}{2!} + M'''(0)\frac{t^3}{3!} + \cdots.$$

The factorial moment-generating function is defined by

$$\eta(t) = E(t^X) = \sum_x t^x f(x) = M[\ln(t)].$$

It can be used to find

$$\mu = E(X) = \eta'(1),$$
$$\sigma^2 = E[X(X-1)] + E(X) - [E(X)]^2 = \eta''(1) + \eta'(1) - [\eta'(1)]^2.$$

Note that $f(x) = P(X = x)$ is the coefficient of $e^{tx}$ in $M(t)$ and the coefficient of $t^x$ in $\eta(t)$.

## EXERCISES

**Purpose:** The exercises illustrate the relation between a moment-generating function, $M(t)$, and the Maclaurin's series for $M(t)$. They also show a method for finding a p.d.f., given either $M(t)$ or $\eta(t)$.

**3.8–1** Let $M(t) = (1-t)^{-1}$, $|t| < 1$, be the moment-generating function of $X$.

(a) Plot the moment-generating function of $X$.

(b) Obtain the Maclaurin series expansion of $M(t)$ (use `taylor`) and verify that

$$M(t) = 1 + t + t^2 + t^3 + \cdots + t^r + \cdots.$$

(c) Let

$$M_r(t) = 1 + t + t^2 + \cdots + t^r.$$

Plot $M(t)$, $M_2(t)$, $M_3(t)$, and $M_4(t)$ on one set of axes. Do $M_2(t)$, $M_3(t)$, and $M_4(t)$ provide good approximations for $M(t)$? For what values of $t$?

(d) What are the values of $E(X)$, $E(X^2)$ and $E(X^3)$?

**3.8–2** Let $M(t) = e^{t^2/2}$ be the moment-generating function of the random variable $X$ of the continuous type.

(a) Graph $y = M(t)$.

(b) Verify that the Maclaurin series expansion of $M(t)$ is

$$M(t) = 1 + \frac{t^2/2}{1!} + \frac{(t^2/2)^2}{2!} + \frac{(t^2/2)^3}{3!} + \cdots.$$

Let

$$M_r(t) = 1 + \cdots + \frac{(t^2/2)^r}{r!}.$$

Graph $M_r(t)$ for $r = 2, 3$, and 4. Do $M_2(t)$, $M_3(t)$, and $M_4(t)$ provide good approximations for $M(t)$? For what values of $t$?

(c) What are the values of $E(X)$, $E(X^2)$ and $E(X^3)$?

**3.8–3** Use *Maple* to derive the moment generating functions for the Bernoulli, binomial, Poisson, geometric, and negative binomial distributions. In each case use the moment-generating function to determine the mean and variance of the distribution.

**3.8–4** Let the distribution of $X$ be $b(5, 1/4)$. Expand the moment-generating function of $X$ and then "strip off" the coefficients of $e^t$.

```
p := 1/4;
n := 5;
M := (1-p+p*E^t)^n;
Mexpand := expand(M);
P := [seq(coeff(Mexpand,E^t,deg),deg = 0 .. n)];
```

What do these coefficients represent? What should their sum be?

**3.8–5** Let $X$ have a geometric distribution with probability of success $p$. Given that the factorial moment-generating function of $X$ is defined by $\eta(t) = pt/(1-(1-p)t)$, expand $\eta(t)$ and then "strip off" the coefficients of $t^x$.

```
eta := p*t/(1-(1-p)*t);
etaseries := series(eta,t,10);
P := [seq(coeff(etaseries,t,deg),deg = 1 .. 10)];
```

What do these coefficients represent?

## Questions and Comments

**3.8–1** If a Maclaurin's series expansion is new for you, show that the Maclaurin's series for $f(x) = e^x$ and $g(x) = \sin\ x$ are given by

(a) $f(x) = e^x = 1 + x + x^2/2! + x^3/3! + \cdots$.

(b) $g(x) = \sin\ x = x - x^3/3! + x^5/5! - x^7/7! + \cdots$.

**3.8–2** Modify Exercises 3.8-4 and 3.8-5 by using other values for the parameters and other discrete distributions. Remember that a moment-generating function is unique for a distribution.

# Chapter 4

# Continuous Distributions

## 4.1 Random Variables of the Continuous Type

The probability density function (p.d.f.) of a random variable $X$ of the continuous type with space $R$, which is an interval or union of intervals, is an integrable function $f$ satisfying the following conditions:

1. $f(x) > 0, \; x \in R,$
2. $\int_R f(x)\,dx = 1,$
3. The probability of the event $X \in A$ is

$$P(X \in A) = \int_A f(x)\,dx.$$

We let $f(x) = 0$ when $x \notin R$.

The (cumulative) distribution function (c.d.f.) of $X$ is defined by

$$F(x) = P(X \leq x) = \int_{-\infty}^{x} f(t)\,dt.$$

The expected value of $X$ or mean of $X$ is

$$\mu = E(X) = \int_{-\infty}^{\infty} x f(x)\,dx.$$

The variance of $X$ is

$$\sigma^2 = \mathrm{Var}(X) = \int_{-\infty}^{\infty} (x-\mu)^2 f(x)\,dx = E(X^2) - \mu^2.$$

The standard deviation of $X$ is

$$\sigma = \sqrt{\mathrm{Var}(X)}.$$

The moment-generating function, if it exists, is

$$M(t) = \int_{-\infty}^{\infty} e^{tx} f(x)\,dx, \quad -h < t < h.$$

## EXERCISES

**Purpose:** The exercises illustrate the relation between a probability density function and the corresponding distribution function. Average distance between randomly selected points is found empirically. Sample skewness is defined and illustrated.

**4.1–1** Sketch the graphs of the following probability density functions and distribution functions associated with these distributions. Also find the mean and the variance for each distribution.

(a) $f(x) = (3/2)x^2, \ -1 \le x \le 1$.

**Remark:** If $f(x)$ is simply defined as the expression `f := (3/2)*x^2`, then its value at $x = -2$ will be $(3/2)(4) = 6$ and not 0. The more precise (and unfortunately somewhat more complicated) definition of $f(x)$ that is given below will make sure that $f(x)$ will be 0 outside of $(-1, 1)$. In particular, this will ensure that the graph of $f(x)$ on intervals that may extend outside of $(-1, 1)$ will be correct. The options given for the `p1` plot are included to prevent *Maple* from connecting the graph at the discontinuities at $x = -1$ and at $x = 1$. The moral to the story is: If you use the simple definition of $f(x)$ (i.e., `f := 3/2*x^2;` or `f := x -> 3/2*x^2;`, then make sure you stay within the $(-1, 1)$ interval.

```
f := proc(x)
if x < -1 then 0
elif 1 < x then 0
else 3/2*x^2
fi
end;
p1 := plot(f,-1.2 .. 1.2,numpoints=500,style=POINT):
F := proc(x)
if x < -1 then 0
elif 1 < x then 1
else int(3/2*t^2,t = -1 .. x)
fi
end;
p2 := plot(F,-1.2 .. 1.2):
plot({p1,p2});
mu := int(3/2*x^3,x = -1 .. 1);
var := int(3/2*(x-mu)^2*x^2,x = -1 .. 1);
```

(b) $f(x) = 1/2, \ -1 \le x \le 1$,

(c) $f(x) = \begin{cases} x+1 & , \ -1 \le x \le 0, \\ 1-x & , \ 0 < x \le 1. \end{cases}$

**4.1–2** Let $X_1$ and $X_2$ denote two random numbers selected from the interval (0,1). Let $D_1 = |X_1 - X_2|$, the random distance between $X_1$ and $X_2$. Illustrate empirically that $E(D_1) = 1/3$ and $\text{Var}(D_1) = 1/18$. In particular, simulate 200 observations of $D_1$. Compare the values of the sample mean and sample variance of the 200 observations with 1/3 and 1/18, respectively. (You will be able to calculate these theoretical values later.)

```
X1 := RNG(200):
X2 := RNG(200):
D1 := [seq(abs(X1[k]-X2[k]),k = 1 .. 200)];
```

**4.1–3** Let $(X_1, Y_1)$ and $(X_2, Y_2)$ denote two points selected at random from the unit square with opposite vertices at (0,0) and (1,1). That is, if $X_1, X_2, Y_1, Y_2$ denote four random numbers selected from the interval (0,1), then

$$D_2 = \sqrt{(X_1 - X_2)^2 + (Y_1 - Y_2)^2}$$

is the distance between these two points. Illustrate empirically that $E(D_2) = 0.5214$. To simulate observations of $D_1$, first simulate observations of $X_1, X_2, Y_1, Y_2$ and notice that `D2[k] := sqrt((X1[k]-X2[k])^2+(Y1[k]-Y2[k])^2);` will be an observation of $D_2$.

**4.1–4** Let $(X_1, Y_1, Z_1)$ and $(X_2, Y_2, Z_2)$ denote two points selected at random from a unit cube with opposite vertices at $(0, 0, 0)$ and $(1, 1, 1)$. That is, if $X_1, X_2, Y_1, Y_2, Z_1$, $Z_2$ denote six random numbers on the interval (0,1) then

$$D_3 = \sqrt{(X_1 - X_2)^2 + (Y_1 - Y_2)^2 + (Z_1 - Z_2)^2}$$

is the distance between these two points. Illustrate empirically that $E(D_3) = 0.6617$.

**Remark**: See *The American Mathematical Monthly*, April, 1978, pages 277–278 for a theoretical solution to this problem.

**4.1–5** Let $C_n = \{(x_1, x_2, \ldots, x_n), -1 \le x_1 \le 1, \ldots, -1 \le x_n \le 1\}$ be a cube in $E^n$, Euclidean $n$-space. If points are selected randomly from $C_n$, on the average how far are these points from the origin? How does the value of $n$ affect this average? That is, what is the limit of this average as $n$ becomes large? And what is the relation between your answers to these questions and your answers to Exercise 2.1-5?

**4.1–6** Skewness, $A_3$, was defined in Section 3.3. Given a set of observations of a random sample, $x_1, x_2, \ldots, x_n$, sample skewness is defined by

$$\alpha_3 = \frac{(1/n)\sum_{i=1}^{n}(x_i - \overline{x})^3}{[(1/n)\sum_{i=1}^{n}(x_i - \overline{x})^2]^{3/2}}.$$

Let the p.d.f. of $X$ be $h(x) = 1,\ 0 \le x \le 1$. Let the p.d.f. of $Y$ be $g(y) = 2 - 2y,\ 0 \le y \le 1$. Let the p.d.f. of $W$ be $f(w) = 3w^2,\ 0 \le w \le 1$.

(a) For each of $X$, $Y$, and $W$, find the values of $\mu$, $\sigma$, and $A_3$.

(b) For each of these distributions, simulate a random sample of size 100 and find the values of the sample mean, sample variance, and sample skewness.

**Hint**: First show that a single observation of each of $X$, $Y$, and $W$ can be simulated using `X := rng()`, `Y := 1-sqrt(1-rng());`, and `W := rng()^(1/3);`. (See Section 4.7.)

(c) Depict a relative frequency histogram of each set of observations with the p.d.f. superimposed. Note the relation between the value of $\alpha_3$ and the shape of the histogram.

## Questions and Comments

**4.1–1** Note very carefully the relation between the p.d.f. and distribution function for each distribution in Exercise 4.1–1. Does the concavity of each distribution function make sense to you?

# 4.2 The Uniform and Exponential Distributions

The p.d.f. of the uniform or rectangular distribution is given by

$$f(x) = \frac{1}{b-a}, \quad a < x < b.$$

We say that $X$ is $U(a,b)$. The mean and variance of $X$ are $\mu = (a+b)/2$ and $\sigma^2 = (b-a)^2/12$, respectively. The distribution function of $X$, for $a < x < b$, is defined by $F(x) = (x-a)/(b-a)$. The p.d.f. and c.d.f. (cumulative distribution function) are defined by `UniformPDF(a .. b,x)` and `UniformCDF(a .. b,x)`. The $p$th percentile of the uniform distribution is given by `UniformP(a .. b,p)`. If $a = 0$ and $b = 1$, observations of $X$ can be simulated using the random number generator. That is, if $X$ is $U(0,1)$, `X := RNG(n);` simulates `n` observations of $X$ while `X := rng();` simulates a single observation of $X$.

Let $X$ be $U(0,1)$. A linear transformation from the interval (0,1) onto the interval $(a,b)$ is given by $y = (b-a)x + a$. Let the random variable Y be defined by $Y = (b-a)X + a$. For $a < y < b$, the distribution function of Y is defined by

$$\begin{aligned} P(Y \le y) &= P((b-a)X + a \le y) \\ &= P\left(X \le \frac{y-a}{b-a}\right) \\ &= \frac{y-a}{b-a}, \end{aligned}$$

the distribution function of the uniform distribution on the interval $(a, b)$, $U(a, b)$. That is `Y := (b-a)*rng()+a;` simulates an observation of the random variable $Y$ that has the uniform distribution $U(a, b)$. `Y := UniformS(a .. b,n);` can be used to simulate `n` observations of $Y$.

The p.d.f.of the exponential distribution with parameter $\theta > 0$ is given by

$$f(x) = (1/\theta)e^{-x/\theta},\ 0 < x < \infty.$$

The distribution function of $X$ for $x > 0$, is

$$F(x) = 1 - e^{-x/\theta}.$$

The mean and variance of $X$ are, respectively, $\mu = \theta$ and $\sigma^2 = \theta^2$. The p.d.f. and c.d.f. for the exponential distribution with mean $\theta$ =t are defined by `ExponentialPDF(t,x)` and `ExponentialCDF(t,x)`. The $p$th percentile of the exponential distribution is given by `ExponentialP(t,p)`.

## EXERCISES

**Purpose:** The simulation of random samples from $U(0, 1)$, $U(a, b)$, and exponential distributions is illustrated. Random numbers are used to approximate values of definite integrals.

**4.2–1** (a) What is the value of the integral $\int_0^1 \sin(\pi x)\,dx$?

(b) Use 500 pairs of random numbers to approximate the value of the integral $\int_0^1 \sin(\pi x)\,dx$. Use either of the following programs.

```
X := RNG(500):
Y := RNG(500):
T := 0:
for k from 1 to 500 do
if Y[k] < evalf(sin(Pi*X[k])) then T := T+1
fi
od;
T;
T/500;
```

```
X := RNG(500):
Y := RNG(500):
Z := [seq(evalf(Y[i]-sin(Pi*X[i])),i = 1 .. 500)]:
ZS := sort(Z):
```

and look at values of `ZS`, e.g., `ZS[310 .. 330];`.

(c) Is your approximation reasonable? If $T$ is equal to the number of pairs of random numbers that fall under the curve $y = \sin(\pi x)$, $0 < x < 1$, how is $T$ distributed?

**4.2–2** (a) Simulate a random sample of size 10, $x_1, x_2, \ldots, x_{10}$, from $U(0,1)$. Use these observations to yield a random sample of size 10, $y_1, y_2, \ldots, y_{10}$, from $U(5,15)$.

(b) State the relation between $\overline{x}$ and $\overline{y}$ and prove this relationship in a general setting.

(c) State the relation between $s_x^2$ and $s_y^2$ and prove it in a general setting.

**Remark:** `U := UniformS(5 .. 15,10);` is a more direct way of getting $y_1, y_2, \ldots, y_{10}$ from $U(5,15)$. You may want to use this in subsequent problems.

**4.2–3** Let $f(x) = e^x$. Use pairs of observations from appropriate uniform distributions to obtain an approximation of $\int_0^1 e^x\,dx$.

**4.2–4** Let $g(x) = (1/\sqrt{2\pi})e^{-x^2/2}$. Use pairs of observations from appropriate uniform distributions to obtain an approximation of $\int_0^2 g(x)\,dx$.

**4.2–5** Let $X$ be $U(a,b)$. Use *Maple* to:

(a) Verify that $\mu = E(X) = (a+b)/2$. (Use `y := UniformPDF(a .. b,x);` followed by `mu := int(x*y,x = a .. b);` and, if necessary, `simplify(mu);`.)

(b) Determine $\sigma^2 = E[(X - \mu^2)]$. (Use `var := int((x-mu)^2*y,x = a .. b);`.)

(c) Derive the moment-generating function, $M(t)$, of $X$.

**4.2–6** (a) Simulate a random sample of size 200 from an exponential distribution with a mean of $\theta = 2$. Why does `X := [seq(-2*log(1-rng()),k = 1 .. 200)];` work? (See Section 4.7.) You may also use `X := ExponentialS(2,200);`.

(b) Find the sample mean and sample variance of your data. Are they close to $\mu = 2$ and $\sigma^2 = 4$?

(c) Depict a relative frequency histogram of your data along with the p.d.f. $f(x) = (1/2)e^{-x/2}$, superimposed. You may use `f := ExponentialPDF(2,x);`.

(d) Depict a relative frequency ogive curve with the theoretical distribution function, $F(x) = 1 - e^{-x/2}$, superimposed. You may use `F := ExponentialCDF(2,x);`.

(e) Depict the empirical distribution function with the theoretical distribution function superimposed.

(f) Make a $q$-$q$ plot of your data with the quantiles for the sample on the $x$-axis and the respective quantiles for the exponential distribution on the $y$-axis. Is this plot linear?

```
X := ExponentialS(2,200):
Y := [seq(ExponentialP(2,k/201),k = 1 .. 200)]:
QQ(X,Y);
```

**4.2–7** Let $X$ be an exponential random variable. Use *Maple* to determine the mean and the variance of $X$.

```
assume(0 < theta);
f := ExponentialPDF(theta,x);
mu := int(x*f,x = 0 .. infinity);
```

**4.2–8** If you are using *Maple* in a windows environment (on a Macintosh computer, Microsoft Windows in DOS, or X-Windows), through animation you can see how the shape of the exponential p.d.f. is affected by $\theta$.

```
with(plots,animate);
f := ExponentialPDF(theta,x);
animate(f,x = 0 .. 8,theta = .1  .. 5);
```

**4.2–9** Let $S$ and $T$ be independent random variables with uniform distributions $U(0,1)$. Show empirically that $X = S + T$ has a triangular distribution.

(a) Generate the sums of 200 pairs of random numbers. For example,

```
S := RNG(200):
T := RNG(200):
X := [seq(S[k]+T[k],k = 1 .. 200)];
```

(b) Plot a relative frequency histogram with the p.d.f. of $X$ superimposed:

$$g(x) = 1 - |x - 1|, \; 0 \le x \le 2.$$

(c) Plot a relative frequency ogive curve with the distribution function of $X$ superimposed.

(d) Plot the empirical distribution function with the theoretical distribution function superimposed.

```
g := x -> 1-abs(x-1);
p1 := Histogram(X,0 .. 2,10):
p2 := plot(g(x),x = 0 .. 2):
plot({p1,p2});
y := int(g(t),t = 0 .. x);
G := x -> y;
p3 := Ogive(X,0 .. 2,10):
p4 := plot(G(x),x = 0 .. 2):
```

```
plot({p3,p4});
p5 := PlotEmpCDF(X,0 .. 2):
plot({p4,p5});
```

## Questions and Comments

**4.2–1** Describe the relation between a relative frequency histogram and a p.d.f.

**4.2–2** Describe the relation between an empirical and a theoretical distribution function.

**4.2–3** In Exercise 4.2–1, would you have more confidence in your answer if you had used more pairs of random numbers? Why?

**4.2–4** Explain how the binomial distribution can be used to say something about the accuracy of estimates of integrals.

# 4.3 The Gamma and Chi-Square Distributions

The p.d.f. for a gamma random variable $X$ is

$$f(x) = \frac{1}{\Gamma(\alpha)\theta^\alpha} x^{\alpha-1} e^{-x/\theta}, \quad 0 < x < \infty.$$

(If $X$ is the waiting time until the $\alpha$th event in a Poisson process with a mean rate of arrivals equal to $\lambda$, then $\theta = 1/\lambda$.) The mean and variance of $X$ are, respectively, $\mu = \alpha\theta$ and $\sigma^2 = \alpha\theta^2$. The distribution function of $X$ is defined, for $x > 0$, by

$$F(x) = \int_0^x f(t)\, dt.$$

The values of the gamma p.d.f. and c.d.f. are given by `GammaPDF(alpha,theta,x);` and `GammaCDF(alpha,theta,x);`, respectively. The $p$th percentile of the gamma distribution is given by `GammaP(alpha,theta,p);`.

Let $X$ have a gamma distribution with $\theta = 2$ and $\alpha = r/2$, where $r$ is a positive integer. The p.d.f. of $X$ is

$$f(x) = \frac{1}{\Gamma(\frac{r}{2})2^{r/2}} x^{(r/2)-1} e^{-x/2}, \quad 0 < x < \infty.$$

We say that $X$ has a chi-square distribution with $r$ degrees of freedom and write $X$ is $\chi^2(r)$. The mean and variance of $X$ are $\mu = \alpha\theta = r$ and $\sigma^2 = \alpha\theta^2 = 2r$. The p.d.f. and c.d.f. for the chi-square distribution are defined by `ChisquarePDF(r,x);` and `ChisquareCDF(r,x);`. The $p$th percentile of the chi-square distribution is given by `ChisquareP(r,p);`.

## EXERCISES

**Purpose:** The exercises illustrate simulations of random samples from the gamma and chi-square distributions. A set of data for a waiting time experiment is included.

**4.3–1** (a) Simulate a random sample of size 200 from a gamma distribution with parameters $\alpha = 5$ and $\theta = 1/2$. Use the fact that a gamma random variable is the sum of $\alpha = 5$ independent exponential random variables, each having a mean of $\theta = 1/2$.

(b) Find the sample mean and sample variance of your observations. Are they approximately equal to the theoretical values?

(c) Depict a relative frequency histogram of your data with the p.d.f. superimposed.

(d) Depict a relative frequency ogive curve with the theoretical distribution function superimposed.

To obtain the 200 gamma observations, you may use

```
for k from 1 to 200 do
XX[k] := 0:
for j from 1 to 5 do
XX[k] := XX[k]-0.5*log(1-rng())
od
od:
X := [seq(XX[k],k = 1 .. 200)]:
```

or

```
A := [seq(ExponentialS(0.5,5),i = 1 .. 200)]:
X := [seq(sum(A[i][j],j = 1 .. 5),i = 1 .. 200)]:
```

For the rest of the problem,

```
p1 := Histogram(X,0 .. 6,10):
y2 := GammaPDF(5,0.5 ,x);
p2 := plot(y2,x = 0 .. 6):
plot({p1,p2});
p3 := Ogive(X,0 .. 6,10):
y3 := GammaCDF(5,0.5 ,x);
p4 := plot(y3,x = 0 .. 6):
plot({p3,p4});
```

(e) You could repeat this exercise using `X := GammaS(5,0.5,200)`.

**4.3–2** The data in Table 4.3–1 are waiting times in seconds to observe 20 $\alpha$-particle emissions of carbon 14 counted by a Geiger counter in a fixed position. Assume that $\lambda = 8$ is the mean number of counts per second and that the number of counts has a Poisson distribution. Plot the empirical distribution function of these data. Superimpose the appropriate gamma distribution function.

| | | | | |
|---|---|---|---|---|
| 2.12 | 2.79 | 1.15 | 1.92 | 2.98 |
| 2.09 | 2.69 | 3.50 | 1.35 | 2.51 |
| 1.31 | 2.59 | 1.93 | 1.75 | 3.12 |
| 2.01 | 1.74 | 2.71 | 1.88 | 2.21 |
| 2.40 | 2.50 | 3.17 | 1.61 | 3.71 |
| 3.05 | 2.34 | 2.90 | 4.17 | 2.72 |

**Table 4.3–1**

**4.3–3** Let $X$ have a gamma distribution with $\alpha = 4$. Depict the p.d.f. of $X$ for $\theta = 2$, $\theta = 4$, and $\theta = 6$ on one set of axes and label each graph.

**4.3–4** Let $X$ have a gamma distribution with $\theta = 4$. Depict the p.d.f. of $X$ for $\alpha = 2$, $\alpha = 4$, and $\alpha = 6$ on one set of axes and label each graph.

**4.3–5** Depict the p.d.f.'s for the chi-square distributions with degrees of freedom 1, 3, and 5 on one set of axes and label each graph.

**4.3–6** If you are using *Maple* in a windows environment (on a Macintosh computer, Microsoft Windows in DOS, or X-Windows), through animation you can see how the shape of the chi-square p.d.f. is affected by $r$.

```
with(plots,animate);
f := ChisquarePDF(r,x);
animate(f,x = 0 .. 12,theta = 0.1  .. 5);
```

**4.3–7** Depict the distribution functions for the chi-square distributions in Exercise 4.3–5.

**4.3–8** Use *Maple* to obtain the means and variances of the chi-square random variables with 1, 3 and 5 degrees of freedom.

**4.3–9** You should have observed in Exercises 4.3–5 and 4.3–6 that the p.d.f. of $\chi^2(1)$ has a dramatically different shape than that of the $\chi^2(3)$ and $\chi^2(5)$ p.d.f.'s. Consider the chi-square p.d.f. as a function of $r$ and $x$ and obtain a three dimensional plot for $0 \le r \le 5$ and $0 \le x \le 10$. Can you tell from this the exact value of $r$ where the change of shape occurs? Can you at least make a reasonable guess?

**4.3–10** Show that the chi-square p.d.f. has a relative maximum if $r > 2$. Where does this maximum occur? What is its value? Answer the question posed in Exercise 4.3–9.

**4.3–11** If $X$ is $\chi^2(9)$, use `ChisquareP(r,p)` to find constants $a$, $b$, $c$, and $d$ so that

(a) $P(X \le a) = 0.05$,

(b) $P(X \le b) = 0.95$,

(c) $P(c \le X \le d) = 0.90$.

(d) Are $c$ and $d$ unique in part (c)?

## Questions and Comments

**4.3–1** Let $X_1$ and $X_2$ be independent and identically distributed exponential random variables with mean $\theta = 2$. How is $Y = X_1 + X_2$ distributed? Can you answer this question empirically?

**4.3–2** Let $X_1, X_2, \ldots, X_5$ be independent and identically distributed exponential random variables with a mean of $\theta = 1/2$. Prove that $W = X_1 + X_2 + \cdots + X_5$ has a gamma distribution with parameters $\alpha = 5$ and $\theta = 1/2$.

**Hint**: Find the moment-generating function of $W$.

**4.3–3** The procedure `GammaS(a,t,n)` can be used to simulate a random sample of size `n` from a gamma distribution with parameters `a` $= \alpha$ and `t` $= \theta$.

**4.3–4** The procedure `ChisquareS(r,n)` can be used to simulate `n` observations of a chi-square random variable with $r$ degrees of freedom.

# 4.4 The Normal Distribution

The p.d.f. of a normally distributed random variable $X$ with mean $\mu$ and variance $\sigma^2$ is

$$f(x) = \frac{1}{\sigma\sqrt{2\pi}} \exp\left[\frac{-(x-\mu)^2}{2\sigma^2}\right], \quad -\infty < x < \infty.$$

We say that $X$ is $N(\mu, \sigma^2)$. If $Z$ is $N(0,1)$, we call $Z$ a standard normal random variable.

The p.d.f. for the $N(\mu, \sigma^2)$ distribution defined by `NormalPDF(mu,sigma^2,x)` and the c.d.f. is defined by `NormalCDF(mu,sigma^2,x)`. The $p$th percentile of the normal distribution can be found using `NormalP(mu,sigma^2,p)`.

## EXERCISES

**Purpose:** The exercises illustrate simulation of normally distributed data, the relation between the $N(0,1)$ and $N(\mu, \sigma^2)$ distributions, the effect of the variance on the distribution, a "test" of a data set for normality, and the calculation of probabilities for normal distributions.

**4.4–1** (a) Use `X := NormalS(0,1,200):` to simulate a random sample of size 200 from a standard normal distribution, $N(0,1)$.

(b) Calculate the sample mean and sample variance. Are they close to $\mu = 0$ and $\sigma^2 = 1$, respectively?

(c) Depict a relative frequency histogram with the $N(0,1)$ p.d.f. superimposed.

(d) Depict a relative frequency ogive curve with the $N(0,1)$ distribution function superimposed.

(f) Make a $q$–$q$ plot with the quantiles for the sample on the $x$-axis and the respective quantiles for the standard normal distribution on the $y$-axis. Is this plot linear?

**Hint**: See Exercise 4.2-6.

**4.4–2** If $Z$ is $N(0,1)$ and $X = \sigma Z + \mu$, then $X$ is $N(\mu, \sigma^2)$. Use this fact to simulate random samples of size 200 from each of the following distributions. For each set of data depict a relative frequency histogram of the data with the superimposed p.d.f.

(a) $N(50,4)$,

(b) $N(50,8)$,

(c) $N(50,16)$.

**Remark**: In part (a), for example, `X := NormalS(50,4,200):` could have been used to obtain a random sample from $N(50,4)$ in a more direct fashion. You may want to use this option in subsequent exercises.

**4.4–3** Often measurements tend to be normally distributed. Select a set of data from your textbook that you think is normally distributed.

(a) Depict a relative frequency histogram with the p.d.f. for a normal distribution superimposed. Use $\overline{x}$ and $s^2$ for the parameters in the p.d.f. Do these data seem to be normally distributed?

(b) Construct a $q$–$q$ plot of the data and the respective quantiles for the standard normal distribution. Is the plot linear? What is the approximate slope of the best fitting line? What is its relation to the standard deviation of the data?

**4.4–4** Use `NormalCDF(0,1,x)` to find the following probabilities when $Z$ is $N(0,1)$:

(a) $P(Z < -1.25)$,

(b) $P(Z > 1.72)$,

(c) $P(-1.47 < Z < 1.98)$.

**4.4–5** Use `NormalCDF(50,64,x)` to find the following probabilities when $X$ is $N(50,64)$:

(a) $P(X < 45.35)$,

(b) $P(X > 59.72)$,

(c) $P(37.41 < X < 63.28)$.

**4.4–6** For a fixed $\mu$, say $\mu = 10$, obtain a 3-dimensional plot of the normal p.d.f. for $0 \le \sigma^2 \le 9$. Describe how the shape of the normal p.d.f. is affected by $\sigma^2$.

**4.4–7** Exercise 4.4–6 provides one way of studying how the shape of the normal p.d.f. is affected by $\sigma^2$. Use animation as an alternative to 3-dimensional plotting to see how the shape of the normal p.d.f. changes as $\sigma^2$ is increased (see Exercise 4.3–6).

## Questions and Comments

**4.4–1** Let $f(x)$ denote the p.d.f. for a normally distributed random variable $X$ where $X$ is $N(\mu, \sigma^2)$.

(a) At what value of $x$ is $f(x)$ a maximum?

(b) What is the value of the maximum?

(c) Between what two values does almost all the probability lie?

**4.4–2** In the future `NormalS(mu,var,n)` can be used to simulate a sample of size `n` from a normal distribution with mean `mu` and variance `var`.

# 4.5 Other Models

If the p.d.f. of $W$ is

$$g(w) = \frac{\alpha w^{\alpha-1}}{\beta^\alpha} \exp\left[-\left(\frac{w}{\beta}\right)^\alpha\right], \quad 0 \le w < \infty,$$

then $W$ has a Weibull distribution.

If the p.d.f. of $Y$ is

$$f(y) = \frac{1}{\sqrt{2\pi}\beta y} \exp\left[-\frac{(\ln y - \alpha)^2}{2\beta^2}\right], \quad 0 \le y < \infty,$$

then $Y$ has a lognormal distribution. The mean and variance of $Y$ are

$$\mu = e^{\alpha+\beta^2/2} \text{ and } \sigma^2 = e^{2\alpha+\beta^2}(e^{\beta^2} - 1).$$

If $X$ is $N(\mu, \sigma^2)$, then $Y = e^X$ has a lognormal distribution with $\alpha = \mu$ and $\beta = \sigma$.

In this section we also look at the Cauchy distribution. Here is one way to introduce this distribution.

A spinner is mounted at the point $(0, 1)$. Let $w$ be the smallest angle between the $y$-axis and the spinner. Assume that $w$ is the value of a random variable $W$ that has a uniform distribution on the interval $(-\pi/2, \pi/2)$. That is, $W$ is $U(-\pi/2, \pi/2)$ and the probability density function for $W$ is

$$f(w) = \frac{1}{\pi}, \quad -\frac{\pi}{2} < w < \frac{\pi}{2}.$$

Let $x$ be a point on the $x$-axis towards which the spinner points. Then the relation between $x$ and $w$ is given by $x = \tan w$ (see figure below).

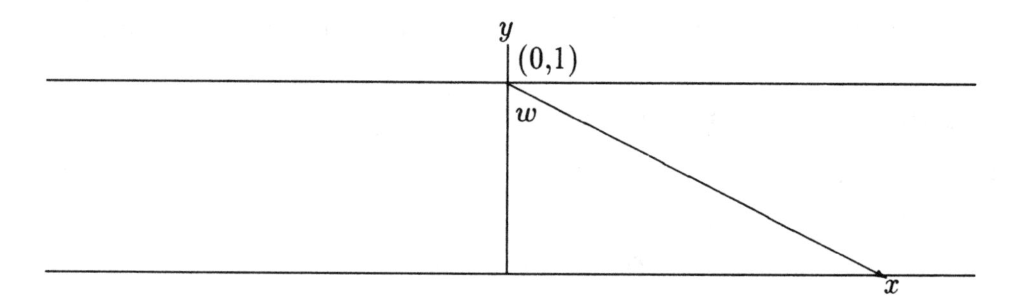


The distribution function of $W$ is given by

$$P(W \le w) = F(w) = \begin{cases} 0, & w < -\pi/2, \\ (1/\pi)(w + \pi/2), & -\pi/2 \le w < \pi/2, \\ 1, & \pi/2 \le w. \end{cases}$$

The distribution function of the random variable $X = \tan W$ is therefore given by

$$\begin{aligned} G(x) &= P(X \le x) \\ &= P(\tan W \le x) \\ &= P(W \le \text{Arctan } x) \\ &= F(\text{Arctan } x) \\ &= \left(\frac{1}{\pi}\right)\left(\text{Arctan } x + \frac{\pi}{2}\right), \quad -\infty < x < \infty. \end{aligned}$$

Hence the probability density function of $X$ is

$$g(x) = G'(x) = \frac{1}{\pi(1 + x^2)}, \quad -\infty < x < \infty.$$

We say that $X$ has a Cauchy distribution.

A more general form of the p.d.f. of the Cauchy distribution is

$$g(x) = \frac{a}{\pi(a^2 + x^2)}, \quad -\infty < x < \infty.$$

## EXERCISES

**Purpose:** The p.d.f.'s for the Weibull and lognormal distributions are graphed for different values of the parameters. Samples are generated from these distributions. Some exercises illustrate that the tail probabilities of the Cauchy distribution are quite large.

**4.5–1** Plot the p.d.f. of a random variable $W$ that has a Weibull distribution with $\beta = 10$ and $\alpha = 1$, $\beta = 10$ and $\alpha = 3$, and $\beta = 10$ and $\alpha = 5$ on one set of axes and label each graph. Note that the p.d.f. is given by `f := WeibullPDF(alpha,beta,x);`.

**4.5–2** Generate a random sample of size 200 from a Weibull distribution with $\alpha = 2$ and $\beta = 10$. Plot a relative frequency histogram with the Weibull p.d.f. superimposed.

**4.5–3** Plot the p.d.f. of a random variable $Y$ that has a lognormal distribution with $\alpha = 0$ and $\beta = 1$ and $\alpha = 2$ and $\beta = 2$ on one set of axes and label each graph.

**4.5–4** (a) Generate a sample of size 200 from a lognormal distribution with $\alpha = 2$ and $\beta = 1/2$.

(b) Plot a relative frequency histogram with the p.d.f. superimposed.

(c) Are the sample mean and sample variance close to the theoretical values?

**4.5–5** Let $X$ have a Cauchy distribution with $a = 1$, and let $x > 0$. Verify that the probabilities given in Table 4.5–1 are correct. In Table 4.5–1, $x = \tan w = \tan \alpha$, where $w$ is in degrees and $\alpha$ is in radians.

| $\alpha$ degrees | $w$ radians | $x = \tan w$ | $P(X > x)$ |
|---|---|---|---|
| 45.000 | 0.7850 | 1.000 | 0.250 |
| 65.000 | 1.1300 | 2.145 | 0.139 |
| 70.000 | 1.2220 | 2.747 | 0.111 |
| 78.690 | 1.3730 | 5.000 | 0.063 |
| 84.290 | 1.4710 | 10.000 | 0.032 |
| 87.710 | 1.5310 | 25.007 | 0.013 |
| 88.850 | 1.5510 | 50.250 | 0.0063 |
| 89.430 | 1.5610 | 100.516 | 0.0032 |
| 89.720 | 1.5660 | 204.626 | 0.0016 |
| 89.886 | 1.5690 | 502.594 | 0.0006 |
| 89.940 | 1.5700 | 954.929 | 0.00033 |
| 89.995 | 1.5707 | 11,459.156 | 0.000028 |

**Table 4.5–1**

**4.5–6** (a) To simulate a random sample of size 400 from the Cauchy distribution, use **one** of the following statements:

```
X := [seq(evalf(tan(Pi*(rng()-1/2))),k = 1 .. 400)]:
X := CauchyS(400):
```

Sort the outcomes and note any outliers. For example, `S := sort(X): S[1..5];` will list the five smallest observations.

(b) Calculate the sample mean and sample variance.

(c) Use `PlotRunningAverage(X);` to depict the running averages of the data. Interpret your graph. What does it say about the fact that the mean of the Cauchy distribution does not exist?

(d) Check some of the probabilities given in Table 4.5–1 empirically. For example check $P(|X| > 25)$, $P(|X| > 50)$, $P(|X| > 100)$.

(e) Plot a relative frequency histogram of your data with the superimposed p.d.f. Because there are often several outliers, it may be useful at this point to set all observations that are less than –5 equal to –5 and those that are greater than 5 equal to 5. You may use `Y := [seq(max(min(X[k],5),-5),k = 1 .. 400)]:`. Compare the histograms for the `X` and the `Y` lists of data.

(f) Graph the ogive and theoretical distribution functions.

(g) Construct box-and-whisker displays for both the `X` and the `Y` lists of data. Describe the important features of these box plots.

## 4.6 Mixed Distributions and Censoring

For most random experiments, distributions of either the discrete type or the continuous type are appropriate. For certain experiments mixed distributions are appropriate.

In this section we shall give examples for which positive probability is assigned to each of certain outcomes that are points and also positive probability is spread over an interval of outcomes, each point of which has zero probability.

### EXERCISES

**Purpose:** The exercises illustrate mixed distributions.

**4.6–1*** Consider the following game. An unbiased coin is flipped. If the outcome is heads the player receives 2 dollars. If the outcome is tails, the player spins a balanced spinner that has a scale from 0 to 1 and receives that fraction of a dollar associated with the point selected by the spinner. Let $X$ denote the amount received.

(a) Show that $\mu = E(X) = 5/4$ and $\sigma^2 = \text{Var}(X) = 29/48$.

(b) Simulate 100 plays of this game.

(c) Find the sample mean and sample variance of these 100 outcomes. Are they close to the theoretical values?

(d) Depict the empirical distribution function.

**4.6–2*** Let $Y_1, Y_2, \ldots, Y_{10}$ denote the outcomes of 10 independent trials of the game described in Exercise 4.6–1. Let

$$\overline{Y} = (1/10)\sum_{i=1}^{10} Y_i.$$

(a) Show that $E(\overline{Y}) = 5/4$ and $\text{Var}(\overline{Y}) = (29/48)/10$.

(b) Let $X = \dfrac{\overline{Y} - 5/4}{\sqrt{(29/48)/10}}$. Show that $E(X) = 0$ and $\text{Var}(X) = 1$.

(c) Simulate 200 observations of $X$, say $X_1, X_2, \ldots, X_{200}$, where each $X_i$ is a function of 10 plays of the game.

(d) Depict a relative frequency histogram of the 200 $X_i$'s.

(e) From the shape of the histogram in part (d), guess the distribution of $X$.

**4.6–3*** The distribution function of $X$ is defined by

$$F(x) = \begin{cases} 0, & x < 0, \\ 1 - (2/3)e^{-x}, & 0 \le x. \end{cases}$$

(a) Find the mean and variance of $X$.

(b) Simulate 100 observations of $X$.

(c) Find the sample mean and sample variance of the 100 observations. Compare these with the theoretical values given in part (a).

(d) Depict the empirical distribution function with the theoretical distribution function superimposed.

## Questions and Comments

**4.6–1** Describe some random experiments for which a mixed distribution is appropriate.

**4.6–2** In Section 5.4 it is shown that $X$, as defined in Exercises 4.6–2, is approximately $N(0, 1)$.

## 4.7 Simulation

In this and previous chapters you have simulated observations from a variety of distributions, both discrete and continuous, without always knowing how such simulations are done. However, we are at a point now where we can describe the basic ideas behind some of the simulation algorithms that you have been using. Since all these algorithms use random numbers, we begin with the generation of random observations from $U(0,1)$.

Linear congruential generators, defined by

$$X_{i+1} \equiv aX_i + c \text{ (modulo } m) \text{ for } i = 0, 1, 2, \ldots,$$

are the most commonly used random number generators on modern computers. In most implementations, the linear congruential generator is simplified to a multiplicative congruential generator by setting $c$ to 0. The choice of $m$ is usually dictated by the word length of the computer and quite frequently, $m$ is set to $2^k$ where $k$ is the number of binary bits that the computer uses to represent positive integers. Many of the 32-bit computers available today have linear congruential generators with $m = 2^{32}$ or $2^{31}$ or $2^{31} - 1$. The advantage of the latter is that it is a prime. Specific choices of $a$, $m$, and $X_0$ determine the period of the generator as well as its statistical qualities.

The multiplicative congruential method for generating random numbers is quite efficient. Furthermore, for certain choices of $a$ and $m$, the generated numbers possess rather good statistical properties. By necessity, numbers generated by a linear congruential scheme have a period after which they repeat themselves. For example, if $m = 15$, the maximum period that can be attained is $2^{15}/4 = 8192$ (see Questions 4.7–1 and 4.7–2 for illustrations of this). Obviously, other things being equal, the longer the period of a generator the better it is.

*Maple* runs on binary computers but it uses its own software code for doing all arithmetic. Thus, there is no inherent advantage for *Maple* to use a power of 2 for $m$. The *Maple* random number generator `rand()` is a multiplicative congruential generator with $m = 999999999989$ (a prime), $a = 427419669081$, and $X_0 = 1$. The period of the generator is $10^{12} - 12$ and it possesses excellent statistical qualities (details can be found in the article by Zaven A. Karian and Rohit Goyal, "Random Number Generation and Testing," *Maple Technical Newsletter*, Vol. 1, No. 1, 1994).

**Simulation of discrete random variables.** Let $X$ be a random variable of the discrete type with support $R = \{b_1, b_2, b_3, \ldots\}$ and $p_i = P(X = b_i) = f(b_i)$. To simulate an observation of $X$, generate a random number $Y = y$, $0 < y < 1$. If $y \le p_1$, let $x = b_1$, if $p_1 < y \le p_1 + p_2$, let $x = b_2$, if $p_1 + p_2 < y \le p_1 + p_2 + p_3$, let $x = b_3$, etc. The procedure `DiscreteS(L,n)` is based on this principle.

**Simulation of continuous random variables.** Let $Y$ have a uniform distribution on the interval $(0,1)$, $U(0,1)$. Let $F(x)$ have the properties of a distribution function of the continuous type. Let $X$ be a function of $Y$ defined by $Y = F(X)$. That is, $X = F^{-1}(Y)$. Then $X$ is a random variable of the continuous type with distribution function $F(x)$.

Thus, if $r_1, r_2, \ldots, r_n$ are random numbers then $x_1 = F^{-1}(r_1),\ x_2 = F^{-1}(r_2), \ldots, x_n = F^{-1}(r_n)$ are random observations of $X$.

As an illustration of the method of generating random observations from a continuous distribution, we apply this method to the exponential distribution. If $X$ is an exponential random variable with a mean $\theta$, the p.d.f. of $X$ is

$$f(x) = (1/\theta)e^{-x/\theta},\ 0 < x < \infty$$

and the distribution function of $X$ for $x > 0$ is

$$F(x) = 1 - e^{-x/\theta}.$$

The mean and variance of $X$ are, respectively, $\mu = \theta$ and $\sigma^2 = \theta^2$.

For $X > 0$, define the random variable $Y$ by

$$Y = F(X) = 1 - e^{-X/\theta}.$$

Then $Y$ is $U(0,1)$. Conversely, if $Y$ is $U(0,1)$, then $X = F^{-1}(Y)$ has an exponential distribution with a mean of $\theta$. That is,

$$X = F^{-1}(Y) = -\theta\ln(1-Y)$$

has an exponential distribution with a mean of $\theta$. Thus, to generate a random sample of size $n$ from this exponential distribution, we obtain $n$ random numbers and apply $F^{-1}$ to each number, `X := [seq(-t*ln(1-rng()),i = 1 .. n)];`, where `t` $= \theta$.

An obvious limitation of this method is its dependence on the availability of $F^{-1}$. In the case of the exponential distribution a simple derivation led us to $F^{-1}(Y) = -\theta\ln(1-Y)$. For other distributions such a derivation could be considerably more complicated; there are also distributions, such as the normal distribution, for which it is impossible to express $F^{-1}$ in closed form. In these cases other methods must be used (Question 4.7–5 describes four methods of dealing with the normal distribution).

## EXERCISES

**Purpose:** The exercises illustrate mixed distributions.

**4.7–1** Let the p.d.f. and distribution function of $X$ be, respectively, $f(x) = 1/(2\sqrt{x})$ and $F(x) = \sqrt{x}$ for $0 < x < 1$. Let Y be $U(0,1)$ and define $X$ by $Y = F(X) = \sqrt{X}$. Then $X = Y^2$ has the given p.d.f. and distribution function.

(a) Simulate 100 observations of $X$.

(b) Plot the empirical and theoretical distribution functions of $X$.

**4.7–2** The p.d.f. of $X$ is $f(x) = 2x,\ 0 < x < 1$.

(a) Find the distribution function of $X$.

(b) Simulate 200 observations of $X$.

(c) Graph the empirical and theoretical distribution functions of $X$.

(d) Depict a relative frequency histogram with the p.d.f. superimposed.

**4.7–3** (a) Let the p.d.f. of $X$ be defined by $f(x) = x/15$, $x = 1, 2, 3, 4, 5$. Find $\mu = E(X)$ and $\sigma^2 = \text{Var}(X)$.

(b) Simulate a random sample of size $n = 100$ from this distribution. You may use `DiscreteS`.

(c) Calculate $\overline{x}$ and $s^2$.

(d) Depict a relative frequency histogram of your data with the superimposed probability histogram.

(e) Depict the empirical distribution function with the superimposed theoretical distribution function.

**4.7–4** Let the p.d.f. of $X$ be defined by $f(x) = (1/2)(2/3)^x$, $x = 1, 2, 3, \ldots$.

(a) Simulate a random sample of size 100 from this distribution.

(b) Is $\overline{x}$ close to $\mu = E(X) = 3$?

(c) Is $s^2$ close to $\sigma^2 = E[(X - \mu)^2] = 6$?

(d) Depict a relative frequency histogram of your data. Superimpose the p.d.f.

## Questions and Comments

**4.7–1** Let $X_{n+1} = 5\,(X_n) \pmod{32}$ and $X_0 = 1$. Find the values of $X_1, X_2, \ldots, X_8$. What does $X_{20}$ equal?

**4.7–2** In Question 4.7–1, let $X_0 = 3$. Find the values of $X_1, X_2, \ldots, X_9$.

**4.7–3** We shall now describe a method for simulating the random sample asked for in Exercise 4.7–4.

Let $F(x)$ be the distribution function of $X$. Let $Y = y$ be the observed value of a random number. For this distribution, if $F(x-1) < y \le F(x)$, then the observed value of the random variable $X$ is $x$, $x = 1, 2, 3, \ldots$. (Plot the distribution function if this is not clear.) Given $y$, our problem is to determine $x$, where $x$ is a positive integer.

Now

$$1 - F(x) = \sum_{k=x+1}^{\infty} \left(\frac{1}{2}\right)\left(\frac{2}{3}\right)^k = \frac{(1/2)(2/3)^{x+1}}{1-(2/3)} = \left(\frac{2}{3}\right)^x.$$

Since $F(x-1) < y \le F(x)$ is equivalent to

$$1 - F(x) \le 1 - y < 1 - F(x-1)$$

we have

$$\left(\frac{2}{3}\right)^x \le 1 - y < \left(\frac{2}{3}\right)^{x-1}.$$

Thus

$$x \ln(2/3) \le \ln(1-y) < (x-1)\ln(2/3)$$

or

$$x \ge \frac{\ln(1-y)}{\ln(2/3)} > x - 1.$$

In this case, let $X = x$. Hence

```
X := [seq(floor(log(1-rng())/log(2/3))+1,i = 1 .. 200)];
```

simulates 200 observations of $X$.

**4.7–4** In addition to its many mathematical features, *Maple* also allows users to program in the *Maple* programming language. In fact, most of the *Maple* procedures that you have been using are written in this language. If you are curious about a particular *Maple* procedure, say the random number generating routine `rand()`, you can actually see its *Maple* code by entering:

```
interface(verboseproc = 2);
eval(rand);
interface(verboseproc = 1);
```

**4.7–5** There are some interesting ways to simulate observations of a standard normal random variable. You may be interested in trying one or more of the following:

(a) `NormalP` can be used to obtain a random sample from the standard normal distribution as follows:

```
Z := [seq(NormalP(0,1,rng()),i = 1 .. 200)]:
```

(b) Let $Y_1, Y_2, \ldots, Y_{12}$ be a random sample of size 12 from $U(0,1)$. Then

$$Z = \frac{\sum_1^{12} Y_i - 12(1/2)}{\sqrt{12(1/12)}} = \sum_{i=1}^{12} Y_i - 6$$

is approximately $N(0,1)$ by the Central Limit Theorem (Section 5.4). That is, the sum of 12 random numbers minus 6 simulates an observation of $Z$, where $Z$ is $N(0,1)$.

```
A := [seq(RNG(12),i = 1 .. 200)];
Z := [seq(sum(A[i][j],j = 1 .. 12),i = 1 .. 200)];
```

(c) Let S and T have independent uniform distributions $U(0,1)$. Let

$$\begin{aligned} X &= \sqrt{-2\ln S}\,\cos(2\pi T), \\ Y &= \sqrt{-2\ln S}\,\sin(2\pi T). \end{aligned}$$

Then $X$ and Y have independent normal distributions $N(0,1)$. (The proof of this depends on an understanding of transformations of continuous random variables. This is the Box-Muller method.) Use

```
Z := [seq(evalf(sqrt(-2*ln(rng()))*cos(2.0*Pi*rng())),
i = 1 .. 200)];
```

or

```
for k from 1 to 200 by 2 do
a := rng():
b := rng():
ZZ[k] := evalf(sqrt(-2*ln(a))*cos(2*Pi*b)):
ZZ[k+1] := evalf(sqrt(-2*ln(a))*sin(2*Pi*b))
od:
Z := [seq(ZZ[k],k = 1 .. 200)];
```

(d) If the random variable $U$ is $U(0,1)$, then the random variable

$$Z = [(U^{0.135} - (1-U)^{0.135}]/0.1975$$

is approximately $N(0,1)$. (*Journal of the American Statistical Association*, September, 1988, page 906.)

```
U := RNG(200):
Z := [seq((U[i]^0.135-(1-U[i])^0.135)/0.1975,
i = 1 .. 200)];
```

# Chapter 5

# Sampling Distribution Theory

## 5.1 Multivariate Distributions

Let $f(x,y)$ be the joint p.d.f. of the discrete type random variables $X$ and $Y$. The marginal p.d.f. of $X$ is defined by

$$f_1(x) = \sum_{y \in R_2} f(x,y), \; x \in R_1,$$

and the marginal p.d.f. of $Y$ is defined by

$$f_2(y) = \sum_{x \in R_1} f(x,y), \; y \in R_2,$$

where $R_1$ and $R_2$ are the spaces of $X$ and $Y$, respectively. The random variables $X$ and $Y$ are independent if and only if

$$f(x,y) \equiv f_1(x) f_2(y) \; \text{ for all } x \in R_1, \; y \in R_2;$$

otherwise $X$ and $Y$ are said to be dependent.

Let $f(x,y)$ denote the joint p.d.f. of the continuous random variables $X$ and $Y$. The marginal p.d.f. of $X$ is given by

$$f_1(x) = \int_{-\infty}^{\infty} f(x,y)\, dy, \quad x \in R_1.$$

The marginal p.d.f. of $Y$ is given by

$$f_2(y) = \int_{-\infty}^{\infty} f(x,y)\, dx, \quad x \in R_2.$$

### EXERCISES

**Purpose:** The exercises illustrate joint distributions of two random variables along with the marginal distributions of these random variables. In addition two random walk problems are given.

**5.1–1** A pair of 8-sided dice is rolled. Let $X$ denote the smaller and $Y$ the larger outcome on the dice.

(a) Define the joint p.d.f. of $X$ and $Y$, the marginal p.d.f. of $X$, and the marginal p.d.f. of $Y$.

(b) Simulate 200 repetitions of this experiment and find the marginal relative frequencies. Explain why this can be done with

```
A := [seq(Die(8,2),i = 1 .. 200)];
B := [seq([min(A[i][1],A[i][2]),max(A[i][1],A[i][2])],
i = 1 .. 200)];
C := MarginalRelFreq(B);
```

(c) Compare the relative frequencies of the outcomes with the respective probabilities. Are they approximately equal?

**5.1–2** Roll nine fair 4-sided dice. Let $X$ equal the number of outcomes that equal 1, and let $Y$ equal the number of outcomes that equal 2 or 3.

(a) Define the joint p.d.f. of $X$ and $Y$, the marginal p.d.f. of $X$, and the marginal p.d.f. of $Y$.

(b) Simulate 200 repetitions of this experiment and find the marginal relative frequencies. Explain why this can be done by

```
A := [seq(Die(4,9),i = 1 .. 200)]:
B := [seq(Freq(A[i],1 .. 4),i = 1 .. 200)]:
C := [seq([B[i][1],B[i][2]+B[i][3]],i = 1 .. 200)]:
D := evalf(MarginalRelFreq(C));
```

(c) Compare the relative frequencies of the outcomes with the respective probabilities. Are they approximately equal?

```
P := [seq(BinomialPDF(9,0.25,x),x = 0 .. 9)];
Q := [seq(BinomialPDF(9,0.50,y),y = 0 .. 9)];
```

(e) Plot the histogram of the observations of $X$ with the p.d.f. of $X$ superimposed.

```
X := [seq(C[i][1],i = 1 .. 200)];
p1 := Histogram(X,-0.5 .. 9.5,10):
p2 := ProbHist(BinomialPDF(9,0.25,t),0 .. 9):
plot({p1,p2});
```

(f) Plot the histogram of the observations of $Y$ with the p.d.f. of $Y$ superimposed.

**5.1–3** A particle starts at $(0,0)$ and moves in one-unit steps with equal probabilities of $1/4$ in each of the four directions—north, south, east, and west. Let $S$ equal the east-west position after $m$ `steps` and let $T$ equal the north-south position after $m$ `steps`. We shall illustrate empirically that each of $X = (S + m)$ and $Y = (T + m)$ has a marginal binomial distribution, $b(2m, 1/2)$.

(a) Use `RandWalk(pn,ps,pe,steps,n);` to simulate 200 ten-steps random walks. Find the values of `X` and `Y`.

```
Position := RandWalk(1/4,1/4,1/4,10,200);
X := [seq(Position[k][1]+10,k = 1 .. 200)];
```

(b) Plot a relative frequency histogram of the 200 observations of `X` with the proper p.d.f. superimposed.

(c) Compare the sample mean and sample variance of the observations of `X` with $\mu = E(X)$ and $\sigma^2 = \text{Var}(X)$.

(d) Repeat parts (b) and (c) for the observations of `Y`.

(e) Define the joint p.d.f. of $S$ and $T$ when $n = 2$. On a two-dimensional graph (on paper), depict the probabilities of the joint p.d.f. and the marginal p.d.f.'s. Are $S$ and $T$ independent? Obtain the marginal relative frequencies and compare them to their corresponding probabilities.

(f) Repeat part (e) for $n = 3$.

(g) Use `GraphRandWalk(pn,ps,pe,steps)` to illustrate 500 steps of this random walk on the monitor screen.

(h) Empirically find the expected number of steps to return to the beginning point.

**5.1–4** A particle starts at $(0,0)$. At each step it moves one unit horizontally and one unit vertically. The horizontal and vertical movements are independent. At the $i$ th step let $p_S$ and $q_S = 1 - p_S$ be the probabilities that the particle moves one unit east and one unit west, respectively. Let $p_T$ and $q_T = 1 - p_T$ be the probabilities that the particle moves one unit north and one unit south, respectively. Let $S$ equal the east-west position and $T$ equal the north-south position after $n$ steps. We shall illustrate empirically that $X = (S + n)/2$ has a binomial distribution $b(n, p_S)$ and $Y = (T + n)/2$ has a binomial distribution $b(n, p_T)$.

(a) Use `RandWalk2(pt,ps,steps,n)` to simulate 200 nine-step random walks with $p_S = 3/4$ and $p_T = 1/4$. For each trial find the values of $X$ and $Y$.

(b) Plot the relative frequency histogram of the 100 observations of $X$ with the proper p.d.f. superimposed.

(c) Compare the sample mean and sample variance of the observations of $X$ with $E(X)$ and $\text{Var}(X)$.

(d) Repeat parts (b) and (c) for the observations of $Y$.

(e) Define the joint p.d.f. of $X$ and $Y$ when $n = 2$, $p_X = p_Y = 1/2$. On a two-dimensional graph (on paper), depict the probabilities of the joint p.d.f. and the marginal p.d.f.'s. Are $X$ and $Y$ independent? Obtain the marginal relative frequencies and compare them to their corresponding probabilities.

(f) Repeat part (e) for $n = 3$.

(g) Use `GraphRandWalk2(ps,pt,steps)` to illustrate 500 steps of this random walk on the monitor screen.

(h) Empirically find the expected number of steps to return to the beginning point.

**5.1–5** Work exercises 2.1–4 and 2.1–5 if you omitted them earlier.

**5.1–6** Let $X_1, X_2, X_3, X_4$ be independent and have distributions that are $U(0,1)$. Illustrate empirically that

(a) $P(X_1 + X_2 \leq 1) = 1/2!$,

(b) $P(X_1 + X_2 + X_3 \leq 1) = 1/3!$,

(c) $P(X_1 + X_2 + X_3 + X_4 \leq 1) = 1/4!$.

(d) Discuss the relationship between this exercise and Exercise 3.3-3 in which the random variable $X$ is the minimum number of random numbers that must be added together so that their sum exceeds one.

## 5.2 Distributions of Sums of Independent Random Variables

Let $X_1, X_2, \ldots, X_n$ be independent random variables with means $\mu_i$ and variances $\sigma_i^2$, $i = 1, 2, \ldots, n$. The mean and variance of

$$Y = \sum_{i=1}^{n} a_i X_i,$$

where $a_1, a_2, \ldots, a_n$ are real constants, are

$$\mu_Y = \sum_{i=1}^{n} a_i \mu_i$$

and

$$\sigma_Y^2 = \sum_{i=1}^{n} a_i^2 \sigma_i^2.$$

If $X_i$ is $N(\mu_i, \sigma_i^2)$, $i = 1, 2, \ldots, n$, then $Y$ is $N(\mu_Y, \sigma_Y^2)$.

Let $X_1, X_2, \ldots, X_n$ be a random sample from the normal distribution $N(\mu, \sigma^2)$. Then

$$\overline{X} = (1/n)\sum_{i=1}^{n} X_i$$

is $N(\mu, \sigma^2/n)$.

Let $X$ and $Y$ be independent discrete random variables with p.d.f.'s $f(k) = P(X = k) = p_k$, $g(k) = P(Y = k) = q_k$, $k = 0, 1, 2, \ldots$, and let $W = X + Y$. The p.d.f.of $W$ is given by

$$\begin{aligned} h(k) &= P(W = k) = \sum_{j=0}^{k} f(j)g(k-j) \\ &= p_0 q_k + p_1 q_{k-1} + \cdots + p_k q_0, \end{aligned}$$

for $k = 0, 1, 2, \ldots$. This relation is called a convolution formula.

## EXERCISES

**Purpose:** The exercises illustrate some of the results empirically. The effect of the sample size on the distribution of $\overline{X}$ is illustrated. Applications of the convolution formula are given for which the computer is useful.

**5.2–1** Let $\overline{X}$ equal the sample mean of a random sample of size $n$ from a normal distribution $N(50, 100)$. Depict the p.d.f. of $\overline{X}$ for $n = 4$, $n = 8$, and $n = 12$ on one set of axes and label each graph.

**5.2–2** Illustrate the result in Exercise 5.2–1 empirically. In particular, for each value of $n$, simulate 100 observations of $\overline{X}$ (`NormalMeanS(50,100,n,100);` may be used). For each $n$ calculate the mean and variance of your observations and plot a relative frequency histogram with the correct p.d.f. superimposed.

**5.2–3** Let $Y_1, Y_2, \ldots, Y_{12}$ denote the outcomes on 12 independent rolls of a fair 4-sided die. Use the convolution formula to find the p.d.f. of $X$ when

(a) $X = Y_1 + Y_2$ (do this by hand and then use *Maple*),

```
Y := [1,1/4,2,1/4,3,1/4,4,1/4];
X := Convolution(Y,Y);
ProbHist(X);
```

(b) $X = Y_1 + Y_2 + Y_3 = (Y_1 + Y_2) + Y_3$,

(c) $X = Y_1 + \cdots + Y_6 = (Y_1 + Y_2 + Y_3) + (Y_4 + Y_5 + Y_6)$,

(d) $X = Y_1 + Y_2 + \cdots + Y_{12}$.

**5.2–4** Verify some of your answers to Exercise 5.2–3 empirically. For example, simulate 100 rolls of twelve 4-sided dice, let $X$ be the sum of the twelve outcomes, and estimate $P(X > 35)$ with $\#(\{x_i : x_i > 35\})/100$.

**5.2–5** (a) Let $X$ equal the sum of 6 independent rolls of a fair 4-sided die. Simulate 100 observations of $X$.

(b) Calculate the sample mean and sample variance of the observations of $X$. Are they approximately equal to $E(X)$ and $\text{Var}(X)$, respectively?

(c) Plot a histogram of the observations of $X$.

(d) Superimpose the probability histogram of $X$ on the histogram of the observations of $X$.

**5.2–6** Four different prizes are put into boxes of cereal. Let $Y$ equal the number of boxes that are purchased to obtain a complete set. Then $Y = X_1 + X_2 + X_3 + X_4$, where $X_i$ has a geometric distribution with $p_i = (5 - i)/4$, $i = 1, 2, 3, 4$.

(a) Find the p.d.f., mean, and variance of $Y$.

**Hint**: You could either use `Convolution` or use the following statements that make use of both the moment-generating function and the factorial moment-generating functions.

```
Mg := p*exp(t)/(1-(1-p)*exp(t));
n := 4;
M := product(subs(p = k/n,Mg),k = 1 .. n);
mu := simplify(subs(t = 0,diff(M,t)));
var := simplify(subs(t = 0,diff(M,t,t))-mu^2);
eta := subs(exp(t) = t,M);
etas := series(eta,t,21);
P := [seq(coeff(etas,t,x),x = 1 .. 20)];
L := [seq(op([x,P[x]]),x = 1 .. 20)];
ProbHist(L);
```

(b) Support your answer to part (a) empirically.

**5.2–7** Let $W$ be $N(27, 9)$, let $Y$ be $N(6, 4)$, and let $W$ and $Y$ be independent. We shall illustrate empirically that $X = W - 2Y$ is $N(15, 25)$.

(a) Simulate a random sample of size 100 from the normal distribution $N(27, 9)$ (`W := NormalS(27,9,100):`), simulate a random sample of size 100 from the normal distribution $N(6, 4)$, (`Y := NormalS(6,4,100):`). To find the values of $X$, let `X := [seq(W[i]-2*Y[i],i = 1 .. 100)]:`.

(b) Find the sample mean and sample variance of the observations of $X$. Are they close to the theoretical values?

(c) Plot a relative frequency histogram of the observations of $X$ with the appropriate p.d.f. superimposed.

**5.2–8** Let $Y$ be $b(16, 1/2)$. Let $X = (Y - np)/\sqrt{npq} = (Y - 8)/2$.

(a) Prove that $E(X) = 0$ and $\text{Var}(X) = 1$.

(b) Simulate 200 observations of $Y$ and the corresponding observations of $X$.

(c) Calculate the values of the sample mean and sample variance of the observed values of $X$. Are they close to 0 and 1, respectively?

(d) Plot a relative frequency histogram of the observations of $X$.

(e) What distribution does the relative frequency histogram of the observations of $X$ look like?

## Questions and Comments

**5.2–1** What effect does the value of $n$ have on the distribution of $\overline{X}$ when sampling from a normal distribution?

**5.2–2** In Section 5.5 we show that $Z = [X - n(5/2)]/\sqrt{n(5/4)}$ is approximately $N(0, 1)$ when $Y$ is the sum of $n$ rolls of a 4-sided die. Use this result with $n = 12$ to approximate $P(X > 35)$ and compare this answer with the empirical answer in Exercise 5.2–4.

**5.2–3** See Section 5.5 for an answer to Exercise 5.2–8(e).

# 5.3 Random Functions Associated with Normal Distributions

Let $X_1, X_2, \ldots, X_n$ be a random sample of size $n$ from a normal distribution, $N(\mu, \sigma^2)$. Let

$$\overline{X} = (1/n)\sum_{i=1}^{n} X_i$$

and

$$S^2 = \frac{1}{n-1}\sum_{i=1}^{n}(X_i - \overline{X})^2.$$

Then

$$\frac{\sum_{i=1}^{n}(X_i - \mu)^2}{\sigma^2} \quad \text{is} \quad \chi^2(n);$$

$$\frac{(n-1)S^2}{\sigma^2} = \frac{\sum_{i=1}^{n}(X_i - \overline{X})^2}{\sigma^2} \quad \text{is} \quad \chi^2(n-1).$$

## EXERCISES

**Purpose:** These results are illustrated empirically. See Section 4.3 for the p.d.f., mean, and variance of the chi-square distribution.

**5.3–1** If $Z$ is $N(0,1)$, then $Z^2$ is $\chi^2(1)$. If $Z_1, Z_2, \ldots, Z_r$ is a random sample of size $r$ from the standard normal distribution, $N(0,1)$, then $W = Z_1^2 + Z_2^2 + \cdots + Z_r^2$ is $\chi^2(r)$.

(a) Use these facts to simulate a random sample of size 100 from a chi-square distribution with $r = 4$ degrees of freedom, $\chi^2(4)$.

(b) Plot a relative frequency histogram with the $\chi^2(4)$ p.d.f. superimposed.

(c) Are the sample mean and sample variance close to the theoretical values?

(d) Plot a relative frequency ogive curve with the $\chi^2(4)$ distribution function superimposed.

**Remark:** `ChisquareS(r,n)` can be used to generate a random sample of size `n` from a $\chi^2(r)$ distribution in a more direct fashion. You may use this in subsequent problems.

**5.3–2** Let $X_1, X_2, \ldots, X_{12}$ be a random sample of size 12 from the normal distribution $N(50, 100)$. Let

$$\overline{X} = (1/12)\sum_{1}^{12} X_i \quad \text{and} \quad S^2 = (1/11)\sum_{1}^{12}(X_i - \overline{X})^2.$$

Illustrate empirically that $E(\overline{X}) = 50$, $\mathrm{Var}(\overline{X}) = 100/12$, $E(S^2) = 100$, and $\mathrm{Var}(S^2) = (100/11)^2(22) = 20000/11$ by simulating 50 means of random samples of size 12 from the normal distribution, $N(50,100)$ (`NormalMeanS(50,100,12,50)` may be used). Calculate the sample mean and sample variance of $\overline{x}_1, \overline{x}_2, \ldots, \overline{x}_{50}$. Now use the procedure `NormalVarianceS(50,100,12,50)` to simulate 50 variances of random samples of size 12 from $N(50,100)$ and calculate the sample mean and sample variance of $s_1^2, s_2^2, \ldots, s_{50}^2$, the 50 observations of $S^2$.

**5.3–3** Let $Y_1, Y_2, Y_3$ be a random sample of size 3 from a normal distribution, $N(37, 16)$. We shall illustrate empirically that

$$S = \sum_{i=1}^{3}(Y_i - 37)^2/16$$

is $\chi^2(3)$ and that

$$T = \sum_{i=1}^{3}(Y_i - \overline{Y})^2/16$$

is $\chi^2(2)$ where

$$\overline{Y} = \frac{1}{3}\sum_{i=1}^{3} Y_i.$$

(a) Generate 100 samples of size 3 from a normal distribution $N(37, 16)$. For each sample of size 3, say $y_1, y_2, y_3$, let

$$s_j = \sum_{i=1}^{3}(y_i - 37)^2/16$$

and let

$$t_j = \sum_{i=1}^{3}(y_i - \overline{y})^2/16$$

where $\overline{y} = (y_1 + y_2 + y_3)/3$, $j = 1, 2, \ldots, 100$.

(b) Find the sample mean and sample variance of $s_1, s_2, \ldots, s_{100}$. Are they close to 3 and 6, respectively?

(c) Plot a relative frequency histogram of the $s$'s with the $\chi^2(3)$ p.d.f. superimposed.

(d) Find the sample mean and sample variance of $t_1, t_2, \ldots, t_{100}$. Are they close to 2 and 4 respectively?

(e) Plot a relative frequency histogram of the $t$'s with the $\chi^2(2)$ p.d.f. superimposed.

**5.3–4** Repeat Exercise 5.3–3 using a sample of size 5 from the normal distribution. Make the appropriate changes in the questions.

**5.3–5** When $\overline{X}$ and $S^2$ are the sample mean and sample variance from a normal distribution, then $\overline{X}$ and $S^2$ are independent. Illustrate this empirically by simulating 100 samples of size $n$ from a standard normal distribution. For each sample, calculate the values of $\overline{x}$ and $s^2$. Use `ScatPlot` to make a scatterplot.

**5.3–6** When $\overline{X}$ and $S^2$ are the sample mean and sample variance from a distribution that is not normal, what can be said about the independence of $\overline{X}$ and $S^2$ and how does the sample size, $n$, affect your answer? Use `ScatPlot` and the technique outlined in Exercise 5.3–5. Then interpret your output.

(a) Sample from the exponential distribution with $\theta = 1$.

(b) Sample from the uniform distribution.

(c) Sample from a distribution of your choice.

**5.3–7** Suppose that the distribution of the weight of a prepackaged "1-pound bag" of carrots is $N(1.18, 0.07^2)$ and the distribution of the weight of a prepackaged "3-pound bag" of carrots is $N(3.22, 0.09^2)$. Selecting bags at random, find the probability that the sum of three 1-pound bags exceeds the weight of one 3-pound bag. Illustrate your answer by graphing the p.d.f. of $Y$, the sum of three 1-pound bags of carrots and the p.d.f. of $W$, the weight of one 3-pound bag of carrots on the same figure. Also graph the p.d.f. of $Y - W$.

**Hint**: First determine the distribution of $Y$, the sum of three 1-pound bags, and then compute $P(Y > W)$, where $W$ is the weight of a 3-pound bag.

## Questions and Comments

**5.3–1** The material in this section is very important for applications in statistical inference.

# 5.4 The Central Limit Theorem

**Central Limit Theorem:** Let $\overline{W}$ be the mean of a random sample $W_1, W_2, \ldots, W_n$ of size $n$ from a distribution with finite mean $\mu$ and finite positive variance $\sigma^2$. Then

$$X_n = \frac{\overline{W} - \mu}{\sigma/\sqrt{n}} = \frac{\sum_{i=1}^{n} W_i - n\mu}{\sqrt{n}\sigma}$$

converges in distribution to a random variable $Z$ that is $N(0, 1)$.

The Central Limit Theorem is used to approximate the distribution function of $X_n$ with the $N(0, 1)$ distribution function, namely,

$$P(X_n \le x) \approx \int_{-\infty}^{x} \frac{1}{\sqrt{2\pi}} e^{-z^2/2}\, dz$$

provided that $n$ is "sufficiently large."

## EXERCISES

**Purpose:** The Central Limit Theorem is illustrated empirically for continuous distributions and for mixed distributions. The effect of the sample size $n$ on the distribution of $X_n$ is investigated to help you decide when $n$ is "sufficiently large" for using a normal approximation. Also investigated is the distribution of $\overline{X}$ when $\mu$ does not exist.

**5.4–1** Let $W_1, W_2, \ldots, W_n$ be a random sample of size $n$ from a uniform distribution, $U(0,1)$. Recall that for this uniform distribution $\mu = 1/2$ and $\sigma^2 = 1/12$. Let $X_n = \sqrt{n}(\overline{W} - 1/2)/\sqrt{1/12}$. We shall investigate whether $X_2$ is approximately $N(0, 1)$.

(a) Simulate 200 means of samples of size 2 from the uniform distribution $U(0,1)$ and calculate the observed values of $X_2$.

(b) Calculate the sample mean and sample variance of the $X_2$'s. Are they close to 0 and 1, respectively?

(c) Plot a relative frequency histogram of the observations of $X_2$ with the $N(0,1)$ p.d.f. superimposed.

(d) Does $X_2$ seem to have a distribution that is approximately $N(0,1)$?

```
n := 2;
A := [seq(RNG(n),k = 1 .. 200)]:
Xn := [seq((Mean(A[k])-0.5)/evalf(sqrt((1/12)/n)),
k = 1 .. 200)]:
Mean(Xn);
Variance(Xn);
h := Histogram(Xn,-3 .. 3,15):
p := plot(NormalPDF(0,1,x),x = -3 .. 3):
plot({p,h});
```

**5.4–2** Repeat Exercise 5.4–1 for $n = 12$ and for other values of $n$ of your choice. When is $n$ "sufficiently large" for $X_n$ to be approximately $N(0,1)$?

**5.4–3** Let $W$ have an exponential distribution with p.d.f. $f(w) = e^{-w}$, $0 < w < \infty$. Note that $\mu = 1$ and $\sigma^2 = 1$. For each of $n = 2, 9, 16$, and 23,

(a) Simulate 200 means of random samples of size $n$ from this exponential distribution. For each sample of size $n$, calculate the value of $X_n = \sqrt{n}(\overline{W} - 1)/\sqrt{1}$. Why does it work to modify lines two and three in the solution for Exercise 5.4-1 as follows?

```
A := ExponentialSumS(1,n,200):
Xn := [seq(evalf((A[k]/n-1)/sqrt(1/n)),k = 1 .. 200)]:
```

(b) Calculate the sample mean and sample variance of the $X_n$'s. Are they close to 0 and 1, respectively?

(c) Plot a relative frequency histogram of the observations of $X_n$ with the $N(0,1)$ p.d.f. superimposed.

(d) How large must $n$ be so that the Central Limit Theorem and the normal distribution provide a "good fit" for means from the exponential distribution?

**5.4–4** Let $W$ have a U-shaped distribution with p.d.f. $f(w) = (3/2)w^2$, $-1 < w < 1$. For this distribution $\mu = 0$ and $\sigma^2 = 3/5$. Explain why an observation of $W$ can be simulated by `W := (2*rng()-1)^(1/3):`. For each of $n = 2, 5, 8$, and 11,

(a) Simulate 200 random samples of size distribution. For each sample of size $n$, calculate the value of $X_n = \sqrt{n}(\overline{W} - 0)/\sqrt{3/5}$.

(b) Calculate the sample mean and sample variance of the $X_n$'s. Are they close to 0 and 1, respectively?

(c) Plot a relative frequency histogram of the observations of $X_n$ with the $N(0,1)$ p.d.f. superimposed.

(d) When is $n$ "sufficiently large" for the normal distribution to provide a "good fit" for means from this U-shaped distribution?

**5.4–5** Let $W$ have a triangular distribution with p.d.f. $f(w) = (w+1)/2, \ -1 < w < 1$. For this distribution $\mu = 1/3$ and $\sigma^2 = 2/9$. Use the method of Section 4.7 to simulate an observation of $W$. For each of $n = 2, 5, 8$, and 11,

(a) Simulate 200 random samples of size $n$ from this triangular distribution. For each sample of size $n$, calculate the value of $X_n = \sqrt{n}(\overline{W} - 1/3)/\sqrt{2/9}$.

(b) Calculate the sample mean and sample variance of the $X_n$'s. Are they close to 0 and 1, respectively?

(c) Plot a relative frequency histogram of the observations of $X_n$ with the $N(0,1)$ p.d.f. superimposed.

(d) When is $n$ "sufficiently large" for the normal distribution to provide a "good fit" for means from this triangular distribution?

**5.4–6** Study the sampling distribution of the means of samples from the normal distribution with mean $\mu = 100$ and variance $\sigma^2 = 144$, $N(100, 144)$.

(a) If $\overline{X}_1, \overline{X}_2, \ldots \overline{X}_{100}$ are 100 means of samples of size $n$, is there a discernible distribution that fits $\overline{X}_1, \overline{X}_2, \ldots \overline{X}_{100}$? Answer this question by generating sets of $\overline{X}_1, \overline{X}_2, \ldots \overline{X}_{100}$ and looking at such things as the average, variance, histogram, and empirical p.d.f. of $\overline{X}_1, \overline{X}_2, \ldots \overline{X}_{100}$.

(b) In a similar way, study the sampling distribution of the medians of samples from $N(100, 144)$.

(c) Do the same for variances.

**5.4–7** * The random variable $W$ has a mixed distribution with distribution function

$$F(w) = \begin{cases} 0, & w < 0, \\ w/2, & 0 \le w < 1, \\ 1/2, & 1 \le w < 2, \\ 1, & 2 \le w, \end{cases}$$

$\mu = 5/4$, $\sigma^2 = 29/48$. (For example, flip a coin. If the outcome is heads, win 2 dollars, if the outcome is tails, spin a balanced spinner with a scale from 0 to 1 and let the payoff equal the outcome on the spinner. See Exercise 4.6-1. For each of $n = 2, 5, 8$, and 11,

(a) Simulate 200 random samples of size $n$ from this mixed distribution. For each sample of size $n$, calculate the value of $X_n = \sqrt{n}(\overline{W} - 5/4)/\sqrt{29/48}$.

(b) Calculate the sample mean and sample variance of the $X_n$'s. Are they close to 0 and 1, respectively?

(c) Plot a relative frequency histogram of the observations of $X_n$ with the $N(0,1)$ p.d.f. superimposed.

(d) When is $n$ "sufficiently large" for the normal distribution to provide a "good fit" for means from this mixed distribution?

**5.4–8** Let $X_1, X_2, \ldots, X_n$ be a random sample of size $n$ from the normal distribution $N(40, 15)$. Let $F_n(\overline{x})$ denote the distribution function of $\overline{X}$ and let the distribution function of $W$ be defined by

$$F(w) = \begin{cases} 0, & w < 40, \\ 1, & 40 \le w. \end{cases}$$

Then $\lim_{n\to\infty} F_n(\overline{x}) = F(\overline{x})$ at each point where F is continuous. We say that the sequence $\{\overline{X}\}$ converges to $W$ *in distribution*. To illustrate this convergence, graph the empirical distribution function of 20 observations of $\overline{X}$ for increasing values of $n$, for example, let $n = 5, 10, 20, 30$.

```
n := 5:
X := NormalMeanS(40,15,n,20);
PlotEmpCDF(X,38 .. 42);
```

**5.4–9** Let $X_1, X_2, \ldots, X_{300}$ be a random sample of size $n = 300$ from the normal distribution, $N(40, 15)$. To illustrate that $\overline{X}$ converges to $\mu = 40$, plot successive sample means of the data.

```
X := NormalS(40,15,300):
p1 := PlotRunningAverage(X,0 .. 300):
p2 := plot(40,x = 0 .. 300):
plot({p1,p2});
```

**5.4–10** Random samples from the Cauchy distribution were simulated in Exercise 4.5-6. In that section it was noted that the Cauchy distribution does not have a mean. What can be said about the sampling distribution of $\overline{X}$?

(a) Simulate 600 observations of a Cauchy random variable. To investigate the limiting behavior of $\overline{X}$ use `PlotRunningAverage`.

(b) Generate 200 observations of $\overline{X}$ for several values of $n$. What can you say about the distribution of $\overline{X}$? Does it have a Cauchy distribution? **Note:** See Exercise 4.5-6(e) for a hint about constructing the histogram.

**5.4–11** Let the p.d.f. of $X$ be $f(x) = 1/x^2, 1 < x < \infty$. Investigate the behavior of the distribution of $\overline{X}$ for several values of $n$.

## Questions and Comments

**5.4–1** When is $n$ "sufficiently large" so that $X_n$ has a distribution that is approximately $N(0,1)$? What properties of the underlying distribution affect your answer?

**5.4–2** A discussion of the chi-square goodness of fit statistic and/or the Kolmogorov-Smirnov goodness of fit statistic would be helpful before working the exercises in this section.

# 5.5 Approximations of Discrete Distributions

Let $Y_1, Y_2, \ldots, Y_n$ be a random sample from a Bernoulli distribution, $b(1,p)$, $0 < p < 1$. Thus $\mu = p$ and $\sigma^2 = p(1-p)$. Let

$$X = \sum_{i=1}^{n} Y_i.$$

Then

$$W_n = \frac{X - np}{\sqrt{np(1-p)}} = \frac{\overline{Y} - p}{\sqrt{p(1-p)/n}}$$

has a limiting normal distribution $N(0,1)$. Thus, if $n$ is "sufficiently large," probabilities for the binomial distribution $b(n,p)$ can be approximated using the normal distribution $N(np, np[1-p])$.

Since the binomial distribution is discrete and the normal distribution is continuous, generally a half unit correction for continuity is made. In particular, if $X$ is $b(n,p)$, for $0 \le k \le n$,

$$\begin{aligned} P(X = k) &= P(k - 1/2 < X < k + 1/2) \\ &\approx P(k - 1/2 < W < k + 1/2) \end{aligned}$$

where $W$ is $N(np, np[1-p])$. A similar type of correction is made for other discrete distributions.

If $Y$ has a Poisson distribution with mean $\lambda$, then $Z = (Y - \lambda)/\sqrt{\lambda}$ is approximately $N(0,1)$ when $\lambda$ is "sufficiently large." That is, the distribution of $Y$ is approximately $N(\lambda, \lambda)$.

If $X$ has a binomial distribution $b(n,p)$, then

$$\frac{(np)^x e^{-np}}{x!} \approx \binom{n}{x} p^x (1-p)^{n-x}, \quad x = 0, 1, 2, \ldots, n.$$

That is, Poisson probabilities can be used to approximate binomial probabilities.

## EXERCISES

**Purpose:** The exercises illustrate applications of approximations of probabilities for discrete distributions.

**5.5–1** (a) Study the shapes of binomial distributions by obtaining probability histograms for $b(n,p)$ for various values of $n$ and $p$. Does it look like the normal distribution could approximate certain binomial distributions? Which ones?

(b) If you wanted to approximate $b(n,p)$ by $N(\mu,\sigma^2)$, what $\mu$ and $\sigma^2$ would you choose? Superimpose the graphs of the $N(\mu,\sigma^2)$ p.d.f.s that you choose on their corresponding $b(n,p)$ probability histograms. How do the approximations look?

(c) Suppose we want to find $P(2 < X \leq 6)$ where $X$ is a $b(15,1/3)$ random variable. By looking at a graph similar to those considered in part (b), you can see that this probability can be approximated by $\int_a^b f(x)\,dx$ where $f(x)$ is the appropriately chosen normal p.d.f. To obtain this approximation, what values of $a$ and $b$ should be used?

(d) Obtain probability histograms for $b(n,p)$ for $p = 0.1$ and $n = 5$, 10, 20, 35 and 50. Does it look like a normal distribution could approximate certain $b(n,0.1)$ distributions? Make some general observations relating $n$ and $p$ to the quality of the approximation.

**5.5–2** (a) Let $X$ have a binomial distribution $b(15,p)$. Simulate a random sample of size 100 from this distribution for $p = 0.1, 0.2, 0.3, 0.4$, and 0.5.

(b) Plot a relative frequency histogram of your observations along with the p.d.f. for the normal distribution $N(15p, 15p(1-p))$ superimposed.

(c) For what values of $np$ does the "fit" appear to be good?

**5.5–3** If you are using *Maple* in a windows environment (on a Macintosh computer, Microsoft Windows in DOS, or X-Windows), you can graphically compare binomial and normal probabilities for various $n$ and $p$ combinations through animation. For example, when $p = 1/8$ the following animates the binomial and corresponding normal p.d.f.s for various values of $n$.

```
with(plots,display);
for i to 10 do
f := BinomialPDF(4*i,1/8,x);
g := NormalPDF(1/2*i,7/16*i,x);
h := ProbHist(f,0 .. 10):
n := plot(g,x = .5  .. 10.5 ):
H.i := plot({n,h})
od:
display([seq(H.i,i = 1 .. 10)],insequence = true);
```

For what values of $n$ and $p$ does the normal distribution provide a good approximation to the binomial?

**5.5–4** (a) Let $X$ have a binomial distribution $b(15, p)$. For $p = 0.1, 0.2, 0.3, 0.4$, and $0.5$, compare the probabilities of each of the 16 possible outcomes with the approximations given by $N(15p, 15p(1-p))$. Recall that `BinomialPDF` gives probabilities for the binomial distribution.

(b) For some choices of $p$ construct a binomial probability histogram with the appropriate normal p.d.f. superimposed. For what values of $np$ does the approximation appear to be good?

**5.5–5** For a fixed $n$, say $n = 20$, obtain a three-dimensional plot of the p.d.f.s of $b(n, p)$ and $N(np, np(1-p))$. For what values of $np$ does the normal approximation to the binomial appear to be good? You may want to "zoom in" to obtain the maximum difference between the two p.d.f.s for several ranges of $p$.

**5.5–6** An unbiased die is rolled 24 times. Let $Y$ be the sum of the 24 resulting values. Approximate $P(Y \geq 86)$ and $P(Y < 86)$

(a) Using the normal distribution,

(b) Empirically by simulating this experiment.

**5.5–7** Let $Y$ have a Poisson distribution with a mean of $\lambda$.

(a) If $\lambda = 30$, use the normal distribution to approximate $P(26 < Y < 37)$ and $P(Y \leq 20)$. Compare these approximations with the probabilities that can be obtained from `PoissonPDF` and/or `PoissonCDF`.

(b) If $\lambda = 16$, use the normal distribution to approximate $P(14 < Y < 21)$ and $P(Y \leq 10)$. Compare these approximations with the probabilities that can be obtained from `PoissonPDF` and/or `PoissonCDF`.

(c) If $\lambda = 9$, use the normal distribution to approximate $P(Y = k)$, $k = 0, 1, \ldots, 19$. Compare these approximations with exact probabilities. Are the approximations good?

(d) Determine the values of $\lambda$ for which you think the normal distribution can be used to approximate Poisson probabilities. Construct a Poisson probability histogram with the $N(\lambda, \lambda)$ p.d.f. superimposed.

**5.5–8** If you are using *Maple* in a windows environment (on a Macintosh computer, Microsoft Windows in DOS, or X-Windows), you can graphically compare Poisson and normal probabilities for various values of $\lambda$ through animation. Modify the animation of Exercise 5.5–3 to study the normal approximation to the Poisson distribution. For what values of $\lambda$ does the normal distribution provide a good approximation to the Poisson?

**5.5–9** Let $X$ have a binomial distribution $b(n, p)$. Let $Y$ have a Poisson distribution with mean $\lambda = np$. Let $W$ have a normal distribution $N(np, np(1-p))$. Compare $P(X = x)$, $x = 0, 1, \ldots, n$, with the approximations given by the distributions of $Y$ and $W$ when

(a) $n = 10$, $p = 0.5$,

(b) $n = 20$, $p = 0.05$,

(c) $n = 50$, $p = 0.1$.

(d) For which values of $n$ and $p$ do the approximations seem to be good?

## Questions and Comments

**5.5–1** A "rule of thumb" states that the normal approximation of binomial probabilities is good if $np \geq 5$ and $n(1-p) \geq 5$. Do you agree with this?

**5.5–2** A "rule of thumb" states that the Poisson approximation of binomial probabilities is good if $n \geq 20$ and $p \leq 0.05$ and is very good if $n \geq 100$ and $p \leq 0.1$. Do you agree with this?

# 5.6 Limiting Moment-Generating Functions

The following theorem is the one that is often used to prove the Central Limit Theorem. It also has applications when, for example, the Poisson distribution is used to approximate binomial probabilities.

**Theorem:** If a sequence of moment-generating functions approaches a certain one, say $M(t)$, then the limit of the corresponding distributions must be the distribution corresponding to $M(t)$.

## EXERCISES

**Purpose:** This theorem is illustrated graphically so that you gain some appreciation of it.

**5.6–1** Plot on the same graph the moment-generating function for the binomial distribution, $b(n, p)$, and the moment-generating function for the Poisson distribution with $\lambda = np$ for different values of $n$ and $p$. For what values of $n$ and $p$ are the moment-generating functions close to each other? Why?

**5.6–2** Let $M(t)$ be the moment-generating function for a distribution that has mean $\mu$. Graph $[M(t/n)]^n$ and $e^{\mu t}$ on the same graph for different moment-generating functions and different values of $n$. Interpret your graphs.

**5.6–3** Illustrate the proof of the Central Limit Theorem graphically.

## 5.7 The $t$ and $F$ Distributions

If $Z$ has a normal distribution $N(0,1)$, $V$ has a chi-square distribution $\chi^2(r)$, and $Z$ and $V$ are independent random variables, then

$$T = \frac{Z}{\sqrt{V/r}}$$

has a $t$ distribution with $r$ degrees of freedom. The p.d.f. of $T$ is

$$f(t) = \frac{\Gamma[(r+1)/2]}{\sqrt{\pi r}\,\Gamma(r/2)(1+t^2/r)^{(r+1)/2}}, \quad -\infty < t < \infty.$$

The mean and variance of $T$ are $\mu = 0$ for $r \geq 2$ and $\sigma^2 = r/(r-2)$ for $r \geq 3$.

Let $X_1, X_2, \ldots, X_n$ be a random sample of size $n$ from a normal distribution $N(\mu, \sigma^2)$. Then

$$T = \frac{\overline{X} - \mu}{S/\sqrt{n}}$$

has a $t$ distribution with $r = n - 1$ degrees of freedom, where $\overline{X}$ is the sample mean and $S$ is the sample standard deviation.

If $U$ has a chi-square distribution $\chi^2(r_1)$, $V$ has a chi-square distribution $\chi^2(r_2)$, and $U$ and $V$ are independent, then

$$F = \frac{U/r_1}{V/r_2}$$

has an $F$ distribution with $r_1$ and $r_2$ degrees of freedom, say $F(r_1, r_2)$. The p.d.f. of $F$ is

$$g(w) = \frac{\Gamma[(r_1+r_2)/2](r_1/r_2)^{r_1/2} w^{(r_1/2)-1}}{\Gamma(r_1/2)\Gamma(r_2/2)(1+r_1 w/r_2)^{(r_1+r_2)/2}}, \quad 0 < w < \infty.$$

The mean and variance of $F$ are

$$\mu = \frac{r_2}{r_2 - 2}$$

and

$$\sigma^2 = \frac{2r_2^2(r_1 + r_2 - 2)}{r_1(r_2-2)^2(r_2-4)}.$$

Let $X_1, X_2, \ldots, X_n$ and $Y_1, Y_2, \ldots, Y_m$ be random samples of sizes $n$ and $m$ from normal distributions $N(\mu_X, \sigma_X^2)$ and $N(\mu_Y, \sigma_Y^2)$, respectively. Then

$$F = \frac{[(n-1)S_X^2]/[\sigma_X^2(n-1)]}{[(m-1)S_Y^2]/[\sigma_Y^2(m-1)]} = \frac{S_X^2/\sigma_X^2}{S_Y^2/\sigma_Y^2}$$

has an $F$ distribution $F(n-1, m-1)$, where $S_X^2$ and $S_Y^2$ are the respective sample variances.

## EXERCISES

**Purpose:** Two methods for simulating random samples from the $t$ and $F$ distributions are given. The first method emphasizes the definition while the second method emphasizes the relation between both the $t$ and $F$ distributions and samples from a normal distribution.

**5.7–1** (a) Plot the p.d.f. for the $t$ distribution with $r = 2$ degrees of freedom using TPDF.

(b) Plot on the same graph the p.d.f. for the normal distribution $N(0, 1)$.

(c) How do these two p.d.f.'s compare?

(d) Compute $E(X)$, $E(X^2)$, and $E(X^3)$ for the two distributions.

**5.7–2** Repeat Exercise 5.7–1 for other values of $r$.

**5.7–3** Let $T$ have a $t$ distribution with $r$ degrees of freedom. Use TP to find $a$ and $b$ so that

(a) $P(T \leq a) = 0.05$ when $r = 7$,

(b) $P(T \leq b) = 0.975$ when $r = 9$,

(c) $P(a \leq T \leq b) = 0.90$ when $r = 6$.

**5.7–4** (a) Generate a random sample of size 100 from the $t$ distribution with $r = 4$ degrees of freedom, say $x_1, x_2, \ldots, x_{100}$. Use the fact that $T = Z/\sqrt{V/4}$, where $Z$ is $N(0, 1)$, $V$ is $\chi^2(4)$, and $Z$ and $V$ are independent random variables.

(b) Find the mean and variance of $x_1, x_2, \ldots, x_{100}$. Are they close to the theoretical mean and variance of $T$?

(c) Plot a relative frequency histogram with the $t$ p.d.f. superimposed.

(d) Plot a relative frequency ogive curve with the $t$ distribution function superimposed.

**5.7–5** If you are using *Maple* in a windows environment (on a Macintosh computer, Microsoft Windows in DOS, or X-Windows), you can graphically compare the p.d.f. of the $t$-distribution to that of $N(0, 1)$ for various degrees of freedom.

```
with(plots,animate);
t := TPDF(nu,x);
n := NormalPDF(0,1,x);
animate({n,t},x = -4 .. 4,nu = 1 .. 16);
```

What happens when the number of degrees of freedom of the $t$-distribution gets large?

**5.7–6** Let $Y_1, Y_2, \ldots, Y_5$ be a random sample of size $n = 5$ from the normal distribution $N(30, 16)$. We shall illustrate empirically that $X = \sqrt{5}(\overline{Y} - 30)/S$ has a $t$ distribution with 4 degrees of freedom.

(a) Generate 200 random samples of size $n = 5$ from the normal distribution $N(30, 16)$. For each sample let $x = \sqrt{5}(\overline{y} - 30)/s$.

(b) Find the sample mean and sample variance of the 100 $x$'s. Are they close to the theoretical mean and theoretical variance of $T$?

(c) Plot the relative frequency histogram of the 100 $x$'s with the $t$ p.d.f. superimposed.

(d) Plot a relative frequency ogive curve with the $t$ distribution function superimposed.

(e) Make a scatter plot of the 200 pairs of points, $(\overline{x}, s)$. Are $\overline{X}$ and $S$ independent?

**5.7–7** In Exercise 5.7–6, take a random sample of size 5 from an exponential distribution with mean 30. Answer the rest of the questions and interpret your output. Try other values of $n$.

**5.7–8** In Exercise 5.7–6, take random samples from a uniform distribution and answer the questions in that exercise.

**5.7–9** Use FPDF to graph the p.d.f. for the $F$ distribution when

(a) $r_1 = 3,\ r_2 = 3$,

(b) $r_1 = 4,\ r_2 = 8$,

(c) $r_1 = 8,\ r_2 = 4$.

**5.7–10** Use FCDF to plot the distribution function for the $F$ distribution when $r_1 = 3$ and $r_2 = 8$.

**5.7–11** Let $F$ have an $F$ distribution with $r_1 = 4$ and $r_2 = 8$ degrees of freedom. Use FP to find constants $a$ and $b$ so that

(a) $P(a \le F \le b) = 0.95$,

(b) $P(a \le F \le b) = 0.97$,

(c) $P(F \le b) = 0.75$.

**5.7–12** (a) Generate a random sample of size 100 from an $F$ distribution with $r_1 = 4$, $r_2 = 8$ degrees of freedom, say $x_1, x_2, \ldots, x_{100}$. Use the fact that $F = (U/4)/(V/8)$, where U is $\chi^2(4)$ and $V$ is $\chi^2(8)$, and $U$ and $V$ are independent.

(b) Find the mean and variance of $x_1, x_2, \ldots, x_{100}$. Are they close to the theoretical mean and variance of $F$?

(c) Plot a relative frequency histogram of your data with the $F$ p.d.f. superimposed.

(d) Plot a relative frequency ogive curve along with the $F$ distribution function.

**5.7–13** Let $W_1, W_2, \ldots, W_5$ and $Y_1, Y_2, \ldots, Y_9$ be random samples of sizes 5 and 9, respectively, from independent normal distributions, $N(85, 25)$ and $N(70, 36)$. We shall illustrate empirically that

$$X = \frac{S_W^2/25}{S_Y^2/36}$$

has an $F$ distribution with $r_1 = 4$, $r_2 = 8$ degrees of freedom.

(a) Generate 100 samples of size 5 from $N(85, 25)$ and 100 samples of size 9 from $N(70, 36)$. From these samples calculate 100 observations of $X$.

(b) Find the sample mean and sample variance of the 100 $x$'s. Are they close to the theoretical mean and variance of $X$?

(c) Plot a relative frequency histogram of your data with the $F$ p.d.f. superimposed.

(d) Plot a relative frequency ogive curve along with the $F$ distribution function.

## Questions and Comments

**5.7–1** In Exercise 5.7–3(c), are $a$ and $b$ unique?

**5.7–2** Did Exercises 5.7–4 and 5.7–6 yield similar results?

**5.7–3** Use random samples from other distributions in Exercise 5.7-6.

**5.7–4** In Exercise 5.7–11(a) and 5.7–11(b), are the constants $a$ and $b$ unique?

**5.7–5** Did Exercises 5.7–12 and 5.7–13 yield similar results?

**5.7–6** In the future `TS` and `FS` may be used to simulate observations of $t$ and $F$ random variables, respectively.

# 5.8 Understanding Variability and Control Charts

Several types of quality control charts are studied in this section.

Suppose that every so often a sample of size $n$ is selected from a production line and measured resulting in measurements $x_1, x_2, \ldots, x_n$. This is repeated say $k$ times. We would expect most of the $k$ sample means, $\overline{x}_1, \overline{x}_2, \ldots, \overline{x}_k$, to plot between $\mu - 3\sigma/\sqrt{n}$ and $\mu + 3\sigma\sqrt{n}$. Since we usually know neither $\mu$ nor $\sigma$, $\mu$ is estimated with $\overline{\overline{x}}$, the average of the $k$ sample means, and $3\sigma\sqrt{n}$ is estimated with $A_3\overline{s}$, where $A_3$ depends on $n$ and $\overline{s}$ is the average of the $k$ sample standard deviations. The formula for $A_3$ is

$$A_3 = \frac{3}{\sqrt{n}}\left[\sqrt{n-1}\,\Gamma\left(\frac{n-1}{2}\right)\right] / \left[\sqrt{2}\Gamma(n/2)\right].$$

A quality control chart for the mean can be plotted using `XbarChart`.

## EXERCISES

**Purpose:** The exercises illustrate the different types of control charts.

**5.8–1** Use the data in your textbook and graph the various control charts. Explain what you see in the graph.

**5.8–2** Find data in a journal, newspaper, or from a local company and construct control charts. Explain what you see in the graph.

**5.8–3** Verify the values of $A_3$ using *Maple.*

```
n := 4;
a3 := 3*sqrt((n-1)/n)*GAMMA((n-1)/2)/sqrt(2)/GAMMA(1/2*n);
evalf(");
```

# 5.9 Transformations of Random Variables

Let $X_1$ and $X_2$ be two independent random variables that have gamma distributions with parameters $(\alpha, \theta)$ and $(\beta, \theta)$, respectively. Then the p.d.f. of $Y = X_1/(X_1 + X_2)$ is

$$g(y) = \frac{\Gamma(\alpha+\beta)}{\Gamma(\alpha)\Gamma(\beta)} y^{\alpha-1}(1-y)^{\beta-1}, \quad 0 < y < 1.$$

This is called a beta p.d.f. with parameters $\alpha$ and $\beta$. For the beta distribution

$$\mu = \frac{\alpha}{\alpha+\beta} \quad \text{and} \quad \sigma^2 = \frac{\alpha\beta}{(\alpha+\beta+1)(\alpha+\beta)^2}.$$

## EXERCISES

**Purpose:** The exercises introduce the Arc Sine Law and illustrate it empirically. Random samples from the beta distribution are simulated.

**5.9–1** * A fair coin is flipped $n = 18$ times. Let $o_k = 1$ if the $k$th trial is heads and let $o_k = -1$ if the $k$th trial is tails. Then $s_k = o_1 + o_2 + \cdots + o_k$ is the cumulative excess of heads over tails at the conclusion of the $k$'th trial. Consider a polygonal path beginning at the origin and connecting in order the 18 points $(k, s_k)$, $k =$ 1,2,...,18. Let $W$ denote the number of line segments from $(k-1, s_{k-1})$ to $(k, s_k)$, $k = 1, 2, , \ldots, 18$, that lie above the $x$-axis. That is, $W$ gives the number of trials for which the number of observed heads is greater than or equal to the number of observed tails. The distribution of $W$ will be investigated. Note that $W$ must be an even integer.

(a) Simulate 200 observations of $W$. Let $X = W/2$. Then $X$ can equal $0, 1, \ldots, 9$.

(b) Plot a relative frequency histogram of the observations of $X$. The p.d.f. of $W$ is given by

$$f(w) = \binom{w}{w/2}\binom{18-w}{(18-w)/2} \cdot \frac{1}{2^{18}}, \quad w = 0, 2, 4, \ldots, 18.$$

Thus the p.d.f. of $X$ is given by

$$g(x) = \binom{2x}{x}\binom{18-2x}{(18-2x)/2} \cdot \frac{1}{2^{18}}, \quad x = 0, 1, \ldots, 9.$$

(c) Let the coin be flipped $n$ times. Then $W/n$ is the proportion of times that the number of heads was greater than or equal to the number of tails. For a fixed $\alpha$, $0 < \alpha < 1$, and large $n$,

$$P(W/n < \alpha) \approx (2/\pi)\text{Arcsin}\,\sqrt{\alpha}.$$

For example, if $n = 18$,

$$P(W < 3) = P(\frac{W}{18} \le \frac{3}{18}) \approx (2/\pi)\text{Arcsin}\sqrt{1/6} = 0.26772.$$

This is to be compared with

$$\begin{aligned} P(W < 3) &= P(W = 0, 2) \\ &= f(0) + f(2) \\ &= 0.18547 + 0.09819 \\ &= 0.28366. \end{aligned}$$

Compare the values of $f(w)$ with the approximations using the inverse sine function when $n = 18$.

(d) Compare the values of $f(w)$ with the approximations using the inverse sine function when $n = 48$.

**Remark:** For a discussion of the discrete arc sine distribution of order $n$, see William Feller: *An Introduction to Probability Theory and Its Applications*, Vol. I, 3rd ed. (New York: John Wiley and Sons, Inc., 1968) pages 78–84. Also see Ruth Heintz: "It's In The Bag," *The Mathematics Teacher*, February 1977, Vol. 70, No. 2, pages 132–136.

**5.9–2** (a) Generate 100 pairs of observations, $(y_1, y_2)$, from two independent gamma distributions with $\alpha = 2$, $\theta = 1/3$ for the distribution of $Y_1$ and $\beta = 7$, $\theta = 1/3$ for the distribution of $Y_2$.

(b) Illustrate empirically that $X = Y_1/(Y_1 + Y_2)$ has a beta distribution, $\alpha = 2$, $\beta = 7$. Plot a relative frequency histogram with beta p.d.f. superimposed.

(c) Compare the sample mean and the sample variance with the theoretical values $\mu = E(X)$ and $\sigma^2 = \text{Var}(X)$.

(d) Illustrate empirically that $X = Y_2/(Y_1 + Y_2)$ has a beta distribution, $\alpha = 7$, $\beta = 2$. Plot a relative frequency histogram with the beta p.d.f. superimposed.

(e) Compare the sample mean and the sample variance with the theoretical values.

**5.9–3** Let $U$ and $V$ have independent chi-square distributions with 4 and 14 degrees of freedom, respectively. Let

$$X = \frac{U}{U+V}$$

Illustrate empirically that $X$ has a beta distribution with $\alpha = 2$ and $\beta = 7$. In particular simulate 100 observations of $X$. Plot a relative frequency histogram with the beta p.d.f. superimposed. Calculate the values of the sample mean and sample variance, comparing them with the theoretical values.

**5.9–4** Let $Z_1$ and $Z_2$ have independent standard normal distributions $N(0,1)$. Let $X = Z_1/Z_2$. Then $X$ has a Cauchy distribution.

(a) Simulate a random sample of size 100 from the Cauchy distribution using this method. Print the outcomes on the screen.

(b) Calculate the sample mean and sample variance.

(c) Check some of the probabilities given in Table 4.5–1 empirically. For example check $P(|X| > 25)$, $P(|X| > 50)$, $P(|X| > 100)$.

(d) Plot a relative frequency histogram of your data. If an observation is greater than 5, group $t$ in the last class. If an observation is less than $-5$, group $t$ in the first class. See Exercise 4.5-6(e). Superimpose the p.d.f.

(e) Depict the ogive relative frequency ogive curve and the theoretical distribution function on the same graph.

**5.9–5** Simulate a random sample of size 1000 from the Cauchy distribution. Find the sample mean and sample variance of the first 100 observations, the first 200 observations, etc. Do these successive sample means and sample variances seem to be converging? Why? Use `PlotRunningAverage` to confirm your answer.

## Questions and Comments

**5.9–1** What is the relation between the outputs for Exercises 5.9–2(b) and 5.9–2(d)?

**5.9–2** Verify theoretically that the mean of the Cauchy distribution does not exist.

# Chapter 6

# Estimation

## 6.1 Properties of Estimators

Let $X_1, X_2, \ldots, X_n$ be a random sample from a distribution that depends on the parameter $\theta$. Let $Y = u(X_1, X_2, \ldots, X_n)$ be a statistic. If $E(Y) = \theta$, then $Y$ is called an unbiased estimator of $\theta$. If, for each positive number $\epsilon$,

$$\lim_{n\to\infty} P(|Y - \theta| \geq \epsilon) = 0$$

or, equivalently,

$$\lim_{n\to\infty} P(|Y - \theta| < \epsilon) = 1,$$

then $Y$ is a consistent estimator of $\theta$. If, among all unbiased estimators, $Y$ is the statistic that minimizes $E[(Y-\theta)^2]$, $Y$ is the minimum variance unbiased estimator of $\theta$.

Let $X_1, X_2, \ldots, X_n$ be a random sample of size $n$ from a normal distribution, $N(\mu, \sigma^2)$. The statistic

$$\overline{X} = \frac{1}{n}\sum_{i=1}^{n} X_i$$

is a minimum variance unbiased estimator of $\mu$ (see Section 6.6)

The statistic

$$S_1 = V = \frac{1}{n}\sum_{i=1}^{n}(X_i - \overline{X})^2$$

is the maximum likelihood estimator of $\sigma^2$. The statistic

$$S_2 = S^2 = \frac{1}{n-1}\sum_{i=1}^{n}(X_i - \overline{X})^2$$

is the minimum variance unbiased estimator of $\sigma^2$.

Both $S_3 = \sqrt{S_1}$ and $S_4 = \sqrt{S_2}$ are biased estimators of $\sigma$. An unbiased estimator of $\sigma$ is

$$S_5 = \frac{\Gamma[(n-1)/2]\sqrt{n/2}}{\Gamma(n/2)}\sqrt{S_1} = \frac{\Gamma[(n-1)/2]\sqrt{(n-1)/2}}{\Gamma(n/2)}\sqrt{S_2}.$$

## EXERCISES

**Purpose:** The exercises illustrate that $\overline{X}$ and $S^2$ are unbiased and consistent estimators of $\mu$ and $\sigma^2$, respectively, when sampling from the normal distribution, $N(\mu, \sigma^2)$. Unbiasedness of estimators of $\sigma^2$ and $\sigma$ is investigated. The variances of competing unbiased estimators are compared.

**6.1–1** Generate 50 random samples of size $n = 5$ from a normal distribution, $N(75, 400)$. For each of these 50 samples calculate the values of $S_1, S_2, S_3, S_4$, and $S_5$ defined above. Print these values for each of the 50 samples. Also calculate and print the values of the means and standard deviations of the 50 $s_1$'s, 50 $s_2$'s, 50 $s_3$'s, 50 $s_4$'s, and 50 $s_5$'s. Are the averages close to the expected values of $S_1, S_2, S_3, S_4$, and $S_5$, respectively?

**6.1–2** Repeat Exercise 6.1-1 using samples of size 10.

**6.1–3** Let $X_1, X_2, X_3$ be a random sample of size 3 from the uniform distribution, $U(\theta - 1/2, \theta + 1/2)$. Let $Y_1 < Y_2 < Y_3$ be the order statistics of this sample. That is $Y_1 = \min\{X_1, X_2, X_3\}$, $Y_2 = \text{median}\{X_1, X_2, X_3\}$, and $Y_3 = \max\{X_1, X_2, X_3\}$. Three possible estimators of $\theta$ are the sample mean

$$W_1 = \overline{X} = \frac{1}{3}\sum_{i=1}^{3} X_i,$$

the sample median, $W_2 = Y_2$, and the midrange, $W_3 = (Y_1 + Y_3)/2$.

(a) Simulate 500 samples of size 3 from $U(\theta - 1/2,\ \theta + 1/2)$ for a particular value of $\theta$, for example $\theta = 1/2$. For each sample, calculate the values of $W_1, W_2$, and $W_3$. Two possible ways to simulate the observations of $W_1, W_2$, and $W_3$ are:

```
Samples := [seq(RNG(3),i = 1 .. 500)]:
W1 := [seq(Mean(Samples[i]),i = 1 .. 500)]:
W2 := [seq(Percentile(Samples[i],.5 ),i = 1 .. 500)]:
W3 := [seq(1/2*Max(Samples[i])+1/2*Min(Samples[i]),
i = 1 .. 500)]:
```

or

```
for k to 500 do
X := RNG(3);
Y := sort(X);
WW1[k] := Mean(X);
WW2[k] := Y[2];
WW3[k] := 1/2*Y[1]+1/2*Y[3]
```

```
od:
W1 := convert(WW1,list):
W2 := convert(WW2,list):
W3 := convert(WW3,list):
```

(b) By comparing the values of the sample means and sample variances of the 500 observations of each of $W_1, W_2$, and $W_3$, which of these statistics seems to be the best estimator of $\theta$?

(c) Compare the histograms of the observations of $W_1, W_2$, and $W_3$. Use 20 classes for each on the interval (0,1).

(d) Verify that $E(W_1) = E(W_2) = E(W_3) = 1/2$ and that $\text{Var}(W_1) = 1/36$, $\text{Var}(W_2) = 1/20$, and $\text{Var}(W_3) = 1/40$ when $\theta = 1/2$. It may be helpful to know that the p.d.f. of the sample mean, $W_1$, is

$$g_1(w) = \begin{cases} (27/2)w^2, & 0 < w < 1/3, \\ (9/2)(-6w^2 + 6w - 1), & 1/3 \le w < 2/3, \\ (27/2)(w-1)^2, & 2/3 \le w < 1. \end{cases}$$

The p.d.f. of the median, $W_2$, is

$$g_2(w) = 6w(1-w), \ 0 < w < 1.$$

The p.d.f. of the midrange, $W_3$, is

$$g_3(w) = \begin{cases} 12w^2, & 0 < w < 1/2, \\ 12(w-1)^2, & 1/2 \le w \le 1. \end{cases}$$

(e) Superimpose each p.d.f. on the respective histogram of the generated data.

**6.1–4** Let $Y_1 < Y_2 < \cdots < Y_5$ be the order statistics (the observations ordered) of the random sample $X_1, X_2, \ldots, X_5$ from the normal distribution $N(\mu, \sigma^2)$. Three unbiased estimators of $\mu$ are the sample mean $W_1 = \overline{X}$, the sample median, $W_2 = Y_3$, and the midrange $W_3 = (Y_1 + Y_5)/2$.

(a) Illustrate empirically that each of $W_1, W_2, W_3$ is an unbiased estimator of $\mu$. In particular, generate 200 samples of size 5 from the normal distribution, $N(70, 20)$. For each sample calculate the values of $W_1, W_2$, and $W_3$. Show that the sample mean of each set of 200 observations is close to 70.

(b) By comparing the sample variances of the observations of $W_1, W_2$, and $W_3$, which of these estimators seems to be the best?

**6.1–5** Let $X_1, X_2, \ldots, X_n$ be a random sample of size $n$ from a normal distribution $N(5, 10)$. Let

$$\overline{X} = \frac{1}{n}\sum_{i=1}^{n} X_i \quad \text{and} \quad S^2 = \frac{1}{n-1}\sum_{i=1}^{n}(X_i - \overline{X})^2.$$

(a) Define the p.d.f. of $\overline{X}$ and give the values of $E(\overline{X})$ and $\text{Var}(\overline{X})$.

(b) For each of $n = 5, 15$, and 25, plot the p.d.f. of $\overline{X}$ on the same graph with $0 < \overline{x} < 10$.

(c) For each of $n = 5, 15$, and 25, generate 100 samples of size $n$ from the normal distribution $N(5, 10)$ and calculate the values of $\overline{x}$ (see part (g)). Calculate the sample mean and sample variance of each sample of 100 $\overline{x}$'s.

(d) Plot a relative frequency histogram of the 100 observations of $\overline{X}$ with its p.d.f. superimposed.

(e) Plot the ogive curve or the empirical distribution function of the 100 observations of $\overline{X}$. Superimpose the distribution function of $\overline{X}$.

(f) Find the values of $E(S^2)$ and $\text{Var}(S^2)$. **Hint:** Recall that $(n-1)S^2/\sigma^2$ is $\chi^2(n-1)$.

(g) For each of the 100 samples of size $n$ generated in part (c) ($n = 5, 15, 25$), calculate the value of $S^2$. Calculate the sample mean and sample variance of each sample of 100 $s^2$'s, comparing these with $E(S^2)$ and $\text{Var}(S^2)$.

(h) Plot a relative frequency histogram of your data. You can define the p.d.f. of $S^2$ and superimpose it over your histogram using, with `var` $= \sigma^2$,

```
f := ChisquarePDF(n-1,(n-1)*x/var)*(n-1)/var:
```

(i) Plot the empirical distribution function of the 100 observations of $S^2$. You can superimpose the theoretical distribution function of $S^2$ using

```
F := ChisquareCDF(n-1,(n-1)*x/var):
```

**6.1–6** Let $X_1, X_2, \ldots, X_n$ be a random sample of size $n$ from a distribution with p.d.f. $f(x;\theta) = \theta\, x^{\theta-1}$, $0 < x < 1$, $0 < \theta < \infty$.

(a) Graph the p.d.f.'s of $X$ for $\theta = 1/4, 4$, and 1.

(b) Show that the method of moments estimator for $\theta$ is $\widetilde{\theta} = \overline{X}/(1 - \overline{X})$.

(c) Illustrate this result empirically. For each of $\theta = 1/4, 4$, and 1, generate 200 observations of $X$. Use `X := [seq(rng()^(1/theta),k = 1 .. 200)]:`. (Why does this work? Empirically show that it works by superimposing the p.d.f. of $X$ on the histogram of the data.) Is $\widetilde{\theta}$ close to $\theta$?

**6.1–7** When sampling from a $N(\mu, \sigma^2)$ distribution, an unbiased estimator of $\sigma$ is $cS$, where

$$c = \frac{\Gamma[(n-1)/2]\sqrt{(n-1)/2}}{\Gamma(n/2)}.$$

Investigate the value of $c$ for different values of $n$. **Note:** `GAMMA(x);` gives the value of $\Gamma(x)$.

## Questions and Comments

**6.1–1** In Exercise 6.1-5, what effect does the sample size $n$ have on the distributions of $\overline{X}$ and $S^2$?

**6.1–2** If you have not already done so, show that $E(S_5) = \sigma$.

**6.1–3** Is $\text{Var}(S_2) = \text{Var}(S^2)$ large or small? Justify your answer using the data that you generated in Exercise 6.1-1 or find $\text{Var}(S^2)$ using the fact that $(n-1)S^2/\sigma^2$ is $\chi^2(n-1)$.

# 6.2 Confidence Intervals for Means

Let $x_1, x_2, \ldots, x_n$ be the observed values of a random sample from a normal distribution, $N(\mu, \sigma^2)$. Let $z_{\alpha/2}$ be a number such that

$$P(Z \geq z_{\alpha/2}) = \alpha/2.$$

Then

$$\left[\overline{x} - z_{\alpha/2}\left(\frac{\sigma}{\sqrt{n}}\right),\ \overline{x} + z_{\alpha/2}\left(\frac{\sigma}{\sqrt{n}}\right)\right]$$

is a $100(1-\alpha)\%$ confidence interval for $\mu$.

If $\sigma^2$ is unknown and $n$ is large, say 30 or greater,

$$\left[\overline{x} - z_{\alpha/2}\left(\frac{s}{\sqrt{n}}\right),\ \overline{x} + z_{\alpha/2}\left(\frac{s}{\sqrt{n}}\right)\right]$$

is an approximate $100(1-\alpha)\%$ confidence interval for $\mu$.

Let $X_1, X_2, \ldots, X_n$ be a random sample of size $n$ from a normal distribution $N(\mu, \sigma^2)$. If $\sigma^2$ is unknown, a $100(1-\alpha)\%$ confidence interval for $\mu$ is

$$\left[\overline{x} - t_{\alpha/2}(n-1)\left(\frac{s}{\sqrt{n}}\right),\ \overline{x} + t_{\alpha/2}(n-1)\left(\frac{s}{\sqrt{n}}\right)\right]$$

where $P[T \geq t_{\alpha/2}(n-1)] = \alpha/2$ and $T$ has $r = n-1$ degrees of freedom.

## EXERCISES

**Purpose:** The exercises illustrate the concept of a confidence interval empirically. A comparison of $z$ and $t$ confidence intervals is given. The effect of the confidence level and the sample size on the length of the interval is demonstrated.

**6.2–1** (a) Generate 50 random samples of size $n = 5$ from a normal distribution, $N(40, 12)$, storing the 50 samples in a list (actually a list of lists), say `L`. For each sample, calculate the endpoints for a 90% confidence interval for $\mu$. You may use `ConfIntMean` to obtain a listing of 50 confidence intervals. To see these confidence intervals, plot them as line segments. You may do this using `ConfIntPlot`. And then to see the relationship of the confidence intervals to the true mean, $\mu = 40$, use `ConfIntPlot(CI,40);`. The following statements will help you with this exercise.

```
L := [seq(NormalS(40,12,5),i = 1 .. 50)]:
CI := ConfIntMean(L,90,12):
ConfIntPlot(CI);
ConfIntPlot(CI,40);
```

(b) Do approximately 90% of the intervals contain the mean? You may use `ConfIntSuc(CI,40);`.

**6.2–2** Repeat Exercise 6.2–1 for $n = 10$.

**6.2–3** Repeat Exercise 6.2–1 for 80% confidence intervals.

**6.2–4** Repeat Exercise 6.2–2 for 80% confidence intervals.

**6.2–5** In this exercise we shall compare confidence intervals for the mean when $\sigma$ is known with the case when $\sigma$ is unknown. Simulate 50 samples of size $n = 5$ from a normal distribution, $N(40, 12)$.

(a) For each sample of size 5, calculate the endpoints for a 90% confidence interval for $\mu$, first assuming that $\sigma^2$ is known. What is the length of these intervals? **Note:** Lengths of intervals (or the average length of the intervals) can be obtained with `ConfIntAvLen(CI)`, where `CI` is a list of lists of the confidence intervals.

(b) For the same 50 samples, calculate the endpoints for 90% confidence intervals for $\mu$ assuming that $\sigma^2$ is unknown. Find the average of the lengths of these intervals. (**Note:** If `ConfIntMean` is invoked with only its first two arguments, then confidence intervals will be computed without any knowledge of $\sigma^2$.) Perhaps the following will help you to better understand how these intervals are actually obtained:

```
n := 5;
m := 50;
z := NormalP(0,1,0.95);
t := TP(n-1,0.95);
randomize();
L := [seq(NormalS(40,12,n),i = 1 .. m)]:
CIz := [seq([evalf(Mean(L[i])-z*sqrt(12/n)),
evalf(Mean(L[i])+z*sqrt(12/n))],i = 1 .. m)]:
CIt := [seq([evalf(Mean(L[i])-t*sqrt(Variance(L[i])/n)),
evalf(Mean(L[i])+t*sqrt(Variance(L[i])/n))],
i = 1 .. m)]:
ConfIntAvLen(CIz);
ConfIntAvLen(CIt);
```

Show that the mean (average) length of the confidence intervals using $t$ and $s$ is

$$E\left[\frac{2\,t_{\alpha/2}(n-1)\,S}{\sqrt{n}}\right] = \frac{2\,t_{\alpha/2}(n-1)}{\sqrt{n}}\frac{\Gamma(n/2)\sqrt{2}}{\Gamma[(n-1)/2)]\sqrt{n-1}}\,\sigma.$$

How do your empirical results compare with this?

(c) For each set of 50 confidence intervals, count the number that contain the mean, $\mu = 40$. Do about 90% of the intervals contain the mean? You may use `ConfIntSuc(CIz,40);` and `ConfIntSuc(CIt,40);`.

(d) Use `ConfIntPlot(CIz,40)` and `ConfIntPlot(CIt,40)` to depict each of the sets of 50 confidence intervals. Graphically compare the two sets of intervals. In particular, what is true about the confidence intervals when $\sigma$ is unknown?

**6.2–6** Repeat Exercise 6.2–5 using 50 samples of size $n = 10$.

**6.2–7** Repeat Exercise 6.2–5 using 80% confidence intervals.

**6.2–8** Repeat Exercise 6.2–6 using 80% confidence intervals.

**6.2–9** The data in Table 6.2–1 give some measurements for 10 male-female pairs of gallinules. The lengths of the culmen are given in millimeters and the weights are given in grams. Let $X$ and $Y$ denote the culmen lengths of the male and female, respectively. Let $U$ and $V$ denote the weights of the male and female, respectively.

(a) Find a 95% confidence interval for $\mu_X$ and for $\mu_Y$.

(b) Find a 95% confidence interval for $\mu_X - \mu_Y$, assuming $X$ and $Y$ are independent random variables. Is this a valid assumption?

(c) If $X$ and $Y$ are dependent random variables, we can take pair-wise difference and find a confidence interval for the mean of these differences. In particular, let $W_k = X_k - Y_k$, $k = 1, 2, \ldots, 10$, the difference of the culmen length for the $k$th pair of birds. Assuming that $W$ is $N(\mu_W, \sigma_W^2)$, find a 95% confidence interval for $\mu_W$.

| | Culmen | Weight | | Culmen | Weight |
|---|---|---|---|---|---|
| M1 | 42.3 | 405 | M6 | 44.0 | 435 |
| F1 | 41.3 | 321 | F6 | 38.3 | 314 |
| M2 | 43.2 | 396 | M7 | 45.5 | 425 |
| F2 | 41.9 | 378 | F7 | 39.1 | 375 |
| M3 | 45.5 | 457 | M8 | 46.2 | 425 |
| F3 | 39.8 | 351 | F8 | 40.9 | 355 |
| M4 | 45.0 | 450 | M9 | 45.5 | 415 |
| F4 | 39.0 | 320 | F9 | 40.9 | 355 |
| M5 | 44.3 | 415 | M10 | 42.8 | 400 |
| F5 | 40.0 | 365 | F10 | 39.5 | 340 |

**Table 6.2-1**

**6.2–10** Let $X$ have the p.d.f. $f(x;\theta) = \theta x^{\theta-1}$, $0 < x < 1$, $0 < \theta < \infty$. Let $X_1, X_2, \ldots, X_n$ denote a random sample of size $n$ from this distribution. Let

$$Y = -\sum_{i=1}^{n} \ln X_i.$$

(a) Show that $E(X) = \theta/(\theta + 1)$.

(b) Show that $\widehat{\theta} = -n/\ln\left(\prod_{i=1}^{n} X_i\right)$ is the maximum likelihood estimator of $\theta$. (See Section 6.6.)

(c) Prove and/or show empirically that the distribution of $W = 2\theta Y$ is $\chi^2(2n)$.

(d) Find a $100(1-\alpha)\%$ confidence interval for $\theta$.

(e*) Illustrate empirically that your answer in part (d) is correct. In particular, simulate 50 90% confidence intervals when $\theta = 3$ and $n = 8$. Illustrate your confidence intervals graphically. What is the mean length of these intervals?

```
a := ChisquareP(16,0.05);
b := ChisquareP(16,0.95);
m := 50:
theta := 3:
for k from 1 to m do
X := [seq(-ln(rng()^(1/theta)),i = 1 .. 8)];
Y := 8*Mean(X);
L[k] := [evalf(a/(2*Y)),evalf(b/(2*Y))]
od:
CI := convert(L,list):
ConfIntPlot(CI,theta);
ConfIntSuc(CI,theta);
ConfIntAvLen(CI);
```

**6.2–11** Let $X_1, X_2, \ldots, X_n$ be a random sample of size $n$ from an exponential distribution having mean $\theta = 15$. Simulate 50 random samples of size $n = 4$ from this distribution. For each sample we shall construct a(n) (approximate) 90% confidence interval for $\theta$ using several possible choices. Recall from the Central Limit Theorem that

$$Z = \frac{\overline{X} - \theta}{\theta/\sqrt{n}}$$

is approximately $N(0,1)$ provided that $n$ *is sufficiently large*. Thus, for a sufficiently large $n$,

$$P\left(-z_{\alpha/2} \leq \frac{\overline{X} - \theta}{\theta/\sqrt{n}} \leq z_{\alpha/2}\right) \approx 1 - \alpha.$$

Solving the inequality yields as endpoints for an approximate $1-\alpha$ confidence interval

$$\overline{x} \pm z_{\alpha/2}\left(\frac{\theta}{\sqrt{n}}\right).$$

Because $\theta$ is unknown, we could replace $\theta$ with $\overline{x}$ yielding the endpoints

$$\overline{x} \pm z_{\alpha/2}\left(\frac{\overline{x}}{\sqrt{n}}\right).$$

Or we could replace $\theta$ with $s$ yielding the endpoints

$$\overline{x} \pm z_{\alpha/2}\left(\frac{s}{\sqrt{n}}\right).$$

And in the latter case we could replace $z_{\alpha/2}$ with $t_{\alpha/2}(n-1)$ yielding

$$\overline{x} \pm t_{\alpha/2}(n-1)\left(\frac{s}{\sqrt{n}}\right).$$

An exact solution of the following inequality for $\theta$,

$$-z_{\alpha/2} \le \frac{\overline{x} - \theta}{\theta/\sqrt{n}} \le z_{\alpha/2}$$

yields as endpoints

$$\left[\frac{\overline{x}}{1 + z_{\alpha/2}/\sqrt{n}}, \frac{\overline{x}}{1 - z_{\alpha/2}/\sqrt{n}}\right].$$

And an additional possibility for a confidence interval for $\theta$ is given in Exercise 6.2-15. Use simulation to decide which of these several intervals is the best. Some things to consider in making your decision are the average lengths of the intervals and the proportion of intervals that contain $\theta = 15$. Remember that we are simulating 90% confidence intervals. Depict the 50 confidence intervals of each type graphically. Do approximately 90% of the intervals contain $\theta = 15$? Why? How do the lengths compare? Which of these several intervals is the best? **Hint:** For each possible choice, calculate and store the endpoints for the confidence intervals in a list of lists `CI` (see Exercise 6.2-10). Use `ConfIntPlot`, `ConfIntSuc`, and `ConfIntAvLen` for your comparisons.

**6.2–12** Repeat Exercise 6.2–11 for $n = 9$.

**6.2–13** Repeat Exercise 6.2–11 for $n = 16$.

**6.2–14** Repeat Exercise 6.2–11 for $n = 25$.

**6.2–15** In Exercise 6.2–11 show that

$$W = \frac{2}{\theta}\sum_{i=1}^{n} X_i$$

is $\chi^2(2n)$ by finding the moment-generating function of $W$. Select $a$ and $b$ using `ChisquareP` so that

$$P\left(a \le \frac{2}{\theta}\sum_{i=1}^{n} X_i \le b\right) = 1 - \alpha.$$

Thus

$$P\left[\frac{2}{b}\sum_{i=1}^{n} X_i \le \theta \le \frac{2}{a}\sum_{i=1}^{n} X_i\right] = 1 - \alpha$$

and

$$\left[\frac{2}{b}\sum_{i=1}^{n} x_i, \frac{2}{a}\sum_{i=1}^{n} x_i\right] = \left[\frac{2n\overline{x}}{b}, \frac{2n\overline{x}}{a}\right]$$

is a $100(1-\alpha)\%$ confidence interval for $\theta$. Use this method to obtain 90% confidence intervals for $\theta$ and compare these intervals with the intervals obtained in Exercises 6.2–11, 6.2–12, 6.2–13, and 6.2–14.

## Questions and Comments

**6.2–1** How does the sample size affect the length of a confidence interval?

**6.2–2** Comment on the lengths of the confidence intervals when the $z$ or the $t$ statistic is used.

**6.2–3** In Exercise 6.2-15, could $a$ and $b$ be selected to minimize the lengths of these intervals? How?

# 6.3 Confidence Intervals for Variances

Let $X_1, X_2, \ldots, X_n$ be a random sample of size $n$ from a normal distribution, $N(\mu, \sigma^2)$. When $\mu$ is unknown, a $100(1-\alpha)\%$ confidence interval for $\sigma^2$ is given by

$$\left[\frac{\sum_{i=1}^n (x_i - \overline{x})^2}{b}, \frac{\sum_{i=1}^n (x_i - \overline{x})^2}{a}\right]$$

where $a$ and $b$ are selected so that

$$P\left(a \le \frac{(n-1)S^2}{\sigma^2} \le b\right) = 1 - \alpha.$$

Usually $a$ and $b$ are selected so that the tail probabilities are equal. Recall that the distribution of $(n-1)S^2/\sigma^2$ is $\chi^2(n-1)$.

A $100(1-\alpha)\%$ confidence interval for $\sigma$ is given by

$$\left[\sqrt{\frac{\sum_{i=1}^n (x_i - \overline{x})^2}{b}}, \sqrt{\frac{\sum_{i=1}^n (x_i - \overline{x})^2}{a}}\right].$$

Let $X_1, X_2, \ldots, X_n$ and $Y_1, Y_2, \ldots, Y_m$ be random samples of sizes $n$ and $m$ from independent normal distributions, $N(\mu_X, \sigma_X^2)$ and $N(\mu_Y, \sigma_Y^2)$, respectively. Let $S_X^2$ and $S_Y^2$ be the respective unbiased estimators of the variances. Then

$$\left[cs_x^2/s_y^2,\ ds_x^2/s_y^2\right]$$

is a $100(1-\alpha)\%$ confidence interval for $\sigma_X^2/\sigma_Y^2$, where $c$ and $d$ are selected using the $F$ distribution, usually with equal tail probabilities so that $c = F_{1-\alpha/2}(m-1, n-1) = 1/F_{\alpha/2}(n-1, m-1)$ and $d = F_{\alpha/2}(m-1, n-1)$.

## EXERCISES

**Purpose:** Confidence intervals for variances are illustrated empirically. Shortest confidence intervals are discussed.

**6.3–1** (a) Simulate 50 random samples of size 10 from the normal distribution, $N(40, 12)$. For each sample of size 10, calculate the endpoints for a 90% confidence interval for $\sigma^2$. Either use `ConfIntVar` or create a list of lists of the confidence intervals, storing them in `CI`. Select $a$ and $b$ to yield equal tail probabilities. Print the endpoints of the confidence intervals.

(b) Find the average lengths of the 50 confidence intervals. You may use `ConfIntAvLen(CI);`.

(c) Calculate the expected value of the length of these confidence intervals and compare the expected value with the observed average length.

(d) Use `ConfIntPlot` to depict the confidence intervals graphically.

(e) What can you say about the variance of the lengths of these confidence intervals?

(f) Use `ConfIntSuc(CI,12)` to calculate the proportion of intervals that contain the variance, $\sigma^2 = 12$.

(g) Find the 90% confidence interval of minimum length for $\sigma^2$ (see Questions and Comments 6.3–1 at the end of this section).

**6.3–2** Repeat Exercise 6.3–1 for $n = 20$.

**6.3–3** Repeat Exercise 6.3–1 for 80% confidence intervals.

**6.3–4** Repeat Exercise 6.3-1, this time finding confidence intervals for the standard deviation, $\sigma$.

**6.3–5** If possible, select $a$ and $b$ so that the confidence intervals have minimum length. (See the Questions and Comments, 6.3–2.)

**6.3–6** Consider the data given in Exercise 6.2–9.

(a) Find a 95% confidence interval for $\sigma_U^2$ and $\sigma_V^2$ and for $\sigma_W^2$.

(b) Find a 95% confidence interval for $\sigma_U^2/\sigma_V^2$, assuming that $U$ and $V$ are independent random variables.

**6.3–7** Empirically examine the behavior of the lengths of the confidence intervals for the ratio of two variances.

## Questions and Comments

**6.3–1** It is possible to select $a$ and $b$ in the confidence interval for $\sigma^2$ so that the resulting interval has the shortest length by solving simultaneously

$$G(b) - G(a) = \int_a^b g(u)\,du = 1 - \alpha$$

and

$$a^2 g(a) = b^2 g(b)$$

where $g(u)$ and $G(u)$ are the p.d.f. and distribution function, respectively, of a chi-square random variable with $r = n - 1$ degrees of freedom. A table of solutions is available in the *American Statistical Association Journal,* September 1959, pages 678-679. More interestingly, obtain your own solutions.

With *Maple* it is possible to solve for $a$ and $b$ when $\alpha$ and $n$ are specified. For example, if $n = 7$ and $\alpha = 0.05$, the following will produce values of $a$ and $b$.

```
n := 7;
alpha := 0.05;
g := ChisquarePDF(n-1,x);
eq1 := int(g,x = a .. b) = 1-alpha;
eq2 := a^2*subs(x = a,g) = b^2*subs(x = b,g);
fsolve({eq2,eq1},{a,b},{a = 0 .. infinity,
b = n-1 .. infinity});
```

**6.3–2** A $100(1-\alpha)\%$ confidence interval for $\sigma$ is given by $\left[\sqrt{(n-1)s^2/b}, \sqrt{(n-1)s^2/a}\,\right]$ where $a$ and $b$ are selected so that

$$P\left(a \le \frac{(n-1)S^2}{\sigma^2} \le b\right) = 1 - \alpha.$$

It is possible to select $a$ and $b$ so that the resulting interval has minimum length by solving simultaneously

$$G(b) - G(a) = \int_a^b g(u)\, du = 1 - \alpha$$

and

$$a^{n/2} e^{-a/2} = b^{n/2} e^{-b/2}$$

where $g(u)$ and $G(u)$ are the p.d.f. and the distribution function, respectively, of a chi-square random variable with $r = n - 1$ degrees of freedom. A table of solutions is given in *Probability and Statistical Inference*, Fourth Edition, Macmillan Publishing Co., Inc. 1993, by Robert V. Hogg and Elliot A. Tanis. You should be able to duplicate that table. A solution is given in an article by Roger Crisman entitled "Shortest Confidence Interval for the Standard Deviation of a Normal Distribution" in the *Journal of Undergraduate Mathematics*, 1975, pages 57-62.

## 6.4 Confidence Intervals for Proportions

Let $X_1, X_2, \ldots, X_n$ be a random sample of size $n$ from a Bernoulli distribution, $b(1, p)$. Then

$$Y = \sum_{i=1}^{n} X_i$$

has a distribution which is $b(n, p)$. By the Central Limit Theorem, the distribution of

$$Z = \frac{Y/n - p}{\sqrt{p(1-p)/n}}$$

is approximately $N(0, 1)$ when $np \geq 5$ and $n(1-p) \geq 5$. Thus

$$P\left[-z_{\alpha/2} \leq \frac{Y/n - p}{\sqrt{p(1-p)/n}} \leq z_{\alpha/2}\right] \approx 1 - \alpha.$$

Letting $\widehat{p} = y/n$, the solution of the inequality provides the following endpoints for an approximate $100(1-\alpha)\%$ confidence interval for $p$:

$$\frac{\widehat{p} + z_{\alpha/2}^2/2n \pm z_{\alpha/2}\sqrt{\widehat{p}(1-\widehat{p})/n + z_{\alpha/2}^2/4n^2}}{1 + z_{\alpha/2}^2/n}$$

where $z_{\alpha/2}$ is selected so that $P(Z \geq z_{\alpha/2}) = \alpha/2$ and $Z$ is $N(0, 1)$. When $n$ is large, this interval is usually replaced with the following approximate $100(1-\alpha)\%$ confidence interval for $p$:

$$\left[\frac{y}{n} - z_{\alpha/2}\sqrt{\frac{(y/n)(1-y/n)}{n}},\ \frac{y}{n} + z_{\alpha/2}\sqrt{\frac{(y/n)(1-y/n)}{n}}\right].$$

## EXERCISES

**Purpose:** The exercises illustrate approximate confidence intervals for $p$.

**6.4–1** (a) Simulate 50 random samples of size $n = 25$ from a Bernoulli distribution, $b(1, 0.85)$. That is, simulate 50 samples of 25 Bernoulli trials. For each sample, find an approximate 90% confidence interval for $p$. You may use the simpler confidence interval. Note that the right endpoint should be at most 1. What percentage of the intervals contain $p = 0.85$?

(b) Use `ConfIntPlot` to depict the confidence intervals.

**6.4–2** Repeat Exercise 6.4–1 with $n = 25$ and $p = 0.75$.

**6.4–3** Repeat Exercises 6.4-1 and 6.4-2 using the longer form of the confidence interval. Which of the two types of intervals is shorter, on the average? (See "Quadratic Confidence Intervals" by Neil C. Schwertman and Larry R. Dion in *The College Mathematics Journal*, Vol. 24, No. 5, November 1993, pages 453-457.)

**6.4–4** Let $X_1, X_2, X_3$ denote a random sample of size three from the standard normal distribution, $N(0,1)$. Call the experiment a success if $X_1 < X_2 < X_3$. Let $p = P(\text{success})$.

(a) Simulate $n = 100$ repetitions of the experiment and count the number of successes. That is, simulate 100 samples of size 3 from the standard normal distribution and count the number of times that $x_1 < x_2 < x_3$. After generating the samples with `S := [seq(NormalS(0,1,3),i = 1 .. 100)]:`, you can do the following

```
count := 0:
for i from 1 to 100 do
if S[i] = sort(S[i]) then count := count+1 fi
od:
count;
```

(b) Give a point estimate of $p$ and give the endpoints for a 95% confidence interval for $p$.

(c) How large a sample is needed so that the estimate of $p$ is within $\epsilon = 0.03$ with 95% confidence? That is, how large a sample is needed so that $y/n \pm \epsilon$, with $\epsilon = 0.03$, is a 95% confidence interval for $p$? (See Section 6.5.)

(d) Repeat parts (a) and (b) using the sample size that was found in part (c).

## Questions and Comments

**6.4–1** Did approximately 90% of the confidence intervals contain $p$ in both Exercise 6.4–1 and Exercise 6.4–3? For which problem would you expect this percentage to be closest to 90%? Why?

# 6.5 Sample Size

How large should the sample size $n$ be so that a confidence interval has the desired confidence level and a given maximum length (or given maximum error)?

Let $X_1, X_2, \ldots, X_n$ be a random sample from a normal distribution with unknown mean $\mu$ and variance $\sigma^2$. If $\sigma^2$ is known, the endpoints for a $100(1-\alpha)\%$ confidence interval for $\mu$ are

$$\overline{x} \pm z_{\alpha/2}\sigma/\sqrt{n}.$$

Let $\epsilon = z_{\alpha/2}\,\sigma/\sqrt{n}$. Thus $2\epsilon$ is the length of the interval. The sample size, $n$, is given by

$$n = \frac{z_{\alpha/2}^2\,\sigma^2}{\epsilon^2}$$

where $z_{\alpha/2}$ and $\epsilon$ are selected to yield the desired confidence level and maximum length of the interval, respectively. If $\sigma^2$ is unknown, an estimate of $\sigma^2$ can be used to find $n$ approximately.

Let $y/n$ denote the proportion of successes in $n$ Bernoulli trials with probability $p$ of success on each trial. The endpoints for an approximate $100(1-\alpha)\%$ confidence interval for $p$ are

$$y/n \pm z_{\alpha/2}\sqrt{\frac{(y/n)(1-y/n)}{n}}.$$

Let

$$\epsilon = z_{\alpha/2}\sqrt{\frac{(y/n)(1-y/n)}{n}}.$$

If it is known that $p$ is close to $p^*$, a known number, we can replace $y/n$ with $p^*$. Solving for $n$ we obtain

$$n = \frac{z_{\alpha/2}^2\,p^*(1-p^*)}{\epsilon^2}.$$

That value of $n$ will provide a confidence interval that has approximately the desired reliability and accuracy. If $p$ is completely unknown, replace $y/n$ with $1/2$. Solving for $n$ we then obtain

$$n = \frac{z_{\alpha/2}^2\,(1/2)(1/2)}{\epsilon^2} = \frac{z_{\alpha/2}^2}{4\epsilon^2}.$$

This value of $n$ will provide a sample size so that the approximate $100(1-\alpha)\%$ confidence interval will be no longer than $2\epsilon$.

## EXERCISES

**Purpose:** Sample sizes are calculated and then it is illustrated empirically that these sample sizes yield confidence intervals having the desired characteristics.

**6.5–1** Suppose that the IQ score for a person selected from a certain group of people is approximately $N(\mu, 100)$. Let $X_1, X_2, \ldots, X_n$ be a random sample of such IQ scores.

(a) How large should $n$ be so that a 90% confidence interval for $\mu$ with maximum error $\epsilon = 3$ is $[\,\overline{x} - 1.645(10)/\sqrt{n},\;\; \overline{x} + 1.645(10)/\sqrt{n}\,]$?

(b) Verify your answer empirically. In particular, simulate 50 random samples of size $n$ from the normal distribution $N(110, 100)$. Do approximately 90% of these samples yield a confidence interval that contains $\mu = 110$?

**6.5–2** How many Bernoulli trials should be observed so that

$$y/n \pm 1.645\sqrt{\frac{(y/n)(1-y/n)}{n}}$$

give the endpoints for an approximate 90% confidence interval for $p$ and $\epsilon = 0.07$ if

(a) It is known that $p$ is close to 0.15?

(b) The value of $p$ is completely unknown?

**6.5–3** (a) Use your answer in Exercise 6.5-2(a) and simulate 50 random samples of $n$ Bernoulli trials with $p = 0.15$. For each sample calculate an approximate 90% confidence interval for $p$. Do approximately 90% of these intervals contain $p = 0.15$?

(b) Use your answer to Exercise 6.5-2(b) and simulate 50 random samples of $n$ Bernoulli trials with $p = 0.15$. For each sample calculate an approximate 90% confidence interval for $p$. Do approximately 90% of these intervals contain $p = 0.15$? Why?

(c) What is the difference between the intervals in parts (a) and (b)?

**6.5–4** Suppose that a gambler bets 1 dollar on red in roulette. Thus the probability of winning 1 dollar is 18/38 and the probability of losing 1 dollar is 20/38. Let $p$ denote the probability that, after placing 100 consecutive 1 dollar bets, the gambler is behind, i.e., has lost more than he has won. We would like to estimate the value of $p$ by simulating this experiment on the computer. Let $X$ equal the amount won or lost after placing 100 bets. Then $p = P(X < 0)$. **Note:** If $W$, the number of wins, is less than the number of losses, then $X < 0$. Since $W$ is $b(100, 18/38)$, $P(X < 0) = P(W < 100 - W) = P(W < 50)$.

(a) How many trials, $n$, must be simulated so that the maximum error of the estimate of $p$ is $\epsilon = 0.05$ with 90% confidence?

(b) Simulate $n$ observations of $X$ where each $X$ equals the amount "won" after placing 100 \$1.00 bets on red in roulette. Give a point estimate of $p = P(X < 0)$.

(c) Graph the histogram of the outcomes of the $n$ simulations of $X$ and superimpose the appropriate normal distribution suggested by the Central Limit Theorem.

## Questions and Comments

**6.5–1** If the set or population from which the sample is taken is infinite, then the variance of $\overline{X}$ is $\sigma^2/n$ where $\sigma^2$ is the variance of the population and $n$ is the sample size. Thus $\epsilon = z_{\alpha/2}\sigma/\sqrt{n}$ is $z_{\alpha/2}$ standard deviations of the sample mean, where $z_{\alpha/2}$ is selected to yield the desired confidence level.

**6.5–2** If the set or population from which the sample is taken is finite and has size $N$, then the variance of $\overline{X}$ is $(\sigma^2/n)(N-n)/(N-1)$, where $\sigma^2$ is the variance of the population and $n$ is the sample size. Again, let $\epsilon = z_{\alpha/2}$ standard deviations of $\overline{X}$, where $z_{\alpha/2}$ is selected from the standard normal table to yield, approximately, the desired confidence level. To determine the appropriate sample size, solve the following equation for $n$:

$$\epsilon = z_{\alpha/2}\left(\frac{\sigma^2}{n}\right)\left(\frac{N-n}{N-1}\right).$$

We first note that

$$n = \left(\frac{z_{\alpha/2}^2\sigma^2}{\epsilon^2}\right)\left(\frac{N-n}{N-1}\right)$$

and if $n$ is small relative to $N$, then $(N-n)/(N-1)$ is close to 1. If this is not the case, we solve for $n$ to obtain

$$n = \frac{z_{\alpha/2}^2\sigma^2 N}{\epsilon^2(N-1) + z_{\alpha/2}^2\sigma^2}.$$

Let

$$m = \frac{z_{\alpha/2}^2\sigma^2}{\epsilon^2}.$$

be a preliminary estimate of $n$. Then

$$n = \frac{m}{1 + (m-1)/N}$$

is the desired sample size.

## 6.6 Maximum Likelihood Estimation

Let $X_1, X_2, \ldots, X_n$ be a random sample of size $n$ from a distribution having p.d.f. $f(x;\theta)$ which depends on a parameter $\theta$. The likelihood function is

$$L(\theta) = \prod_{i=1}^{n} f(x_i;\theta).$$

If $u(x_1, x_2, \ldots, x_n)$ is the value of $\theta$ that maximizes $L(\theta)$, then $\widehat{\theta} = u(X_1, X_2, \ldots, X_n)$ is called the maximum likelihood estimator of $\theta$.

## EXERCISES

**Purpose:** The exercises illustrate maximum likelihood estimation for several distributions.

**6.6–1** Let $X$ have a Poisson distribution.

(a) Show that the maximum likelihood estimator of $\lambda$ is $\widehat{\lambda} = \overline{X}$ and $E(\overline{X}) = \lambda$. Furthermore, if

$$S^2 = \frac{1}{n-1}\sum_{i=1}^{n}(X_i - \overline{X})^2,$$

show that $E(S^2) = \lambda$.

(b) Simulate 100 random samples of size $n = 10$ from a Poisson distribution with a mean of $\lambda = 4.5$. For each of these 100 samples, calculate $\overline{X}$ and $S^2$.

(c) Calculate the sample mean and sample variance of the 100 $\overline{x}$'s and the 100 $s^2$'s.

(d) Which of the statistics, $\overline{X}$ or $S^2$, seems to be a better estimator of $\lambda$? Why?

**6.6–2** Let $X_1, X_2, \ldots, X_n$ be a random sample of size $n$ from a distribution with p.d.f. $f(x;\theta) = (1/\theta)x^{(1-\theta)/\theta}$, $0 < x < 1$. Note that the distribution function of $X$ is $F(x) = x^{1/\theta}$, $0 < x < 1$, and the mean is $E(X) = 1/(1+\theta)$.

(a) Graph the p.d.f. of $X$ for $\theta = 1$, $\theta = 1/4$, and $\theta = 4$.

(b) Show that the maximum likelihood estimator of $\theta$ is

$$U = \widehat{\theta} = -\frac{1}{n}\sum_{i=1}^{n}\ln X_i = -\ln (X_1 X_2 \cdots X_n)^{1/n}.$$

(c) If we let $\overline{X} = 1/(1+\theta)$, show that $V = \widetilde{\theta} = (1-\overline{X})/\overline{X}$ is the method of moments estimator of $\theta$.

(d) We shall empirically compare $U$ and $V$ as estimators of $\theta$. Let $\theta = 4$ and generate 100 samples of size 10 from this distribution. For each of the 100 samples, calculate the values of $U$ and $V$. These are estimates of $\theta$. Compare the sample means and sample variances of these two sets of 100 estimates. Which of the estimators, $U$ or $V$, seems to be better? How do their histograms compare?

(e) Repeat part (d) with $\theta = 1/4$.

**6.6–3** Let $X_1, X_2, \ldots, X_{10}$ be a random sample of size 10 from a geometric distribution.

(a) Show that the maximum likelihood estimator of $p$ is $\widehat{p} = 1/\overline{X}$.

(b) Simulate 50 samples of size 10 from a geometric distribution for which $p = 1/10$. For each sample calculate the value of $\widehat{p}$. Is $\widehat{p}$ always close to $1/10$?

(c) Find the average of the 50 observations of $\widehat{p}$. Is this average close to $1/10$?

**6.6–4** Let $X$ have a double exponential distribution with p.d.f. $f(x;\theta) = (1/2)e^{-|x-\theta|}$, $-\infty < x < \infty, -\infty < \theta < \infty$.

(a) Show that the maximum likelihood estimator of $\theta$ is $\widehat{m}$, the sample median.

(b) Show that the method of moments estimator of $\theta$ is $\overline{X}$.

(c) Simulate 100 samples of size 3 from this double exponential distribution for some selected value of $\theta$, e.g., $\theta = 0$. For each sample of size 3, calculate the value of $\widehat{m}$ and the value of $\overline{x}$. For a specified `theta`, the simulation can be done by

```
Y := RNG(100):
for k from 1 to 100 do
if Y[k] < 0.5  then S[k] := theta+ln(2*Y[k])
else S[k] := theta-ln(2-2*Y[k])
fi
od:
X := convert(S,list);
```

Derive the inverse of the distribution function of $X$ and explain why this works.

(d) Calculate the sample mean and sample variance of the sets of 100 observations of $\widehat{m}$ and 100 observations of $\overline{x}$.

(e) Which of the statistics, $\widehat{m}$ or $\overline{x}$, would you recommend as the better statistic to use for estimating $\theta$? Use your empirical results in part (d) to support your answer.

(f) Plot relative frequency histograms of the observations of $\overline{X}$ and $\widehat{m}$. When $n = 3$, the p.d.f. of the median (see Section 10.1) is given by

```
g := 1.5*exp(-2*abs(y))*(1-0.5*exp(-abs(y)));
```

Superimpose this p.d.f. on the histogram of the medians.

## Questions and Comments

**6.6–1** For the Poisson distribution, is it possible to give an estimate of the variance based on a single observation? Why?

**6.6–2** If $U$ and $V$ are two unbiased statistics, that is, $E(U) = E(V) = \theta$, and if $\mathrm{Var}(U) < \mathrm{Var}(V)$, which of the two statistics would be a better estimator for $\theta$? Why?

## 6.7 Asymptotic Distributions of MLE's

Let $X$ have the p.d.f. $f(x;\theta)$ and let $\widehat{\theta}$ be the maximum likelihood estimator of $\theta$. Then $\widehat{\theta}$ has an approximate normal distribution with mean $\theta$ and standard deviation

$$1/\sqrt{-nE\{\partial^2[\ln f(x;\theta)]/\partial\theta^2\}}.$$

### EXERCISES

**Purpose:** The exercise illustrates the asymptotic distribution of $\widehat{\theta}$.

**6.7–1** Let $X_1, X_2, \ldots, X_n$ be a random sample of size $n$ from a normal distribution with known mean, $\mu$, and variance $\theta = \sigma^2$.

(a) Show that $\widehat{\theta} = \sum_{i=1}^{n}(X_i - \mu)^2/n$ is the maximum likelihood estimator of $\theta$.

(b) Show that the approximate distribution of $\widehat{\theta}$ is normal with mean $\theta = \sigma^2$, and standard deviation $\theta/\sqrt{n/2}$.

(c) Illustrate part (b) empirically. In particular simulate 200 observations of size $n$ from the standard normal distribution. Determine the values of $n$ for which

$$\frac{\widehat{\theta} - \theta}{\theta/\sqrt{n/2}} = \frac{\widehat{\theta} - 1}{1/\sqrt{n/2}}$$

is approximately $N(0,1)$ by superimposing the $N(0,1)$ p.d.f. over the histogram of the observations.

(d) Show, either theoretically or empirically, that the distribution of $n\widehat{\theta}/\theta$ is $\chi^2(n)$.

## 6.8 Chebyshev's Inequality

Chebyshev's inequality states that if a random variable $X$ has a mean $\mu$ and finite variance $\sigma^2$, then for every $k \geq 1$,

$$P(|X - \mu| \geq k\sigma) \leq \frac{1}{k^2}.$$

If we let $\epsilon = k\sigma$, this becomes

$$P(|X - \mu| \geq \epsilon) \leq \frac{\sigma^2}{\epsilon^2}.$$

Each of these inequalities can be written in terms of complementary events, namely,

$$P(|X - \mu| < k\sigma) \geq 1 - \frac{1}{k^2}$$

and

$$P(|X - \mu| < \epsilon) \geq 1 - \frac{\sigma^2}{\epsilon^2}.$$

In terms of the empirical distribution, given a collection of numbers $x_1, x_2, \ldots, x_n$ with empirical mean $\overline{x}$ and empirical variance $v$, for every $k \geq 1$,

$$\frac{\#\{x_i : |x_i - \overline{x}| \geq k\sqrt{v}\}}{n} \leq \frac{1}{k^2}$$

or, equivalently,

$$\frac{\#\{x_i : |x_i - \overline{x}| < k\sqrt{v}\}}{n} \geq 1 - \frac{1}{k^2}.$$

## EXERCISES

**Purpose:** The exercises investigate how well Chebyshev's inequality bounds the given probabilities.

**6.8–1** For each of the following distributions, simulate 500 observations and compare the proportions of observations that lie within 1, 2, and 3 standard deviations of the mean with the bound that is given by Chebyshev's inequality. What conclusions can you draw?

(a) $Z$ is $N(0,1)$.

(b) $X$ has an exponential distribution with mean $\theta = 1$.

(c) $Y$ has a binomial distribution, $b(n,p)$. Try different combinations of $n$ and $p$.

(d) $W = Z_1/\sqrt{(Z_1^2 + Z_2^2)/2}$ where $Z_1$ and $Z_2$ are independent $N(0,1)$ random variables, the mean of $W$ is 0, and the variance of $W$ is 1.

**6.8–2** If $X$ is a random variable with mean $\mu$ and standard deviation $\sigma$, then the mean and standard deviation of $W = (X - \mu)/\sigma$ are 0 and 1. In terms of $W$, the second form of Chebyshev's inequality becomes

$$P(|W - 0| < k) \geq 1 - \frac{1}{k^2}$$

for $k \geq 1$. If there exists a continuous random variable $W$ such that

$$P(|W - 0| < k) = 1 - \frac{1}{k^2}$$

and the distribution is symmetric about the origin, then the distribution function, $F(w)$, of $W$ satisfies

$$F(w) - F(-w) = 1 - \frac{1}{w^2}, \quad w \geq 1,$$

and

$$F(-w) = 1 - F(w).$$

(a) Show that the assumptions on the distribution function imply that the p.d.f. of $W$ is

$$f(w) = \begin{cases} \dfrac{1}{|w|^3}, & |w| > 1 \\ 0, & |w| \le 1. \end{cases}$$

(b) Show that $F^{-1}(x)$ can be defined by

$$F^{-1}(x) = \begin{cases} \dfrac{-1}{\sqrt{2x}}, & 0 < x < 1/2, \\ \dfrac{1}{\sqrt{2(1-x)}}, & 1/2 \le x < 1. \end{cases}$$

(c) Define $F^{-1}(y)$ as a *Maple* function, where $0 < y < 1$,

```
FInv := proc(y)
if y < 1/2 then -1/sqrt(2*y)
else 1/sqrt(2-2*y)
fi
end;
```

Now use it to simulate observations of $W$ with `Y := RNG(500):` followed by `W := [seq(FInv(Y[i]),i = 1 .. 500)]:`. Superimpose the p.d.f. over the histogram of the observations to compare the theoretical and empirical distributions.

(d) The distribution of $W$ was developed so that

$$P(|W - 0| < k) = 1 - \frac{1}{k^2}.$$

Does your empirical data support this equality? In particular, for $k = 2, 3, 4, 5, 6$, is the proportion of observations that are within $k$ of $\mu = 0$ approximately equal to $1/k^2$?

(e) Does the distribution of $W$ satisfy the hypotheses of Chebyshev's inequality? In particular, find the mean and variance of $W$.

(f) For your simulated data, find the proportion of observations that are within $k$ (for $k = 2, 3, 4$) standard deviations ($\sqrt{v}$) of $\overline{x}$. How do these proportions compare with Chebyshev's guarantee?

**6.8–3** Continuing with the last example, let $g(w) = c/|w|^3$, $1 \le |w| \le d$.

(a) Show that in order for $g(w)$ to be a p.d.f., it must be true that $c = d^2/(d^2-1)$.

(b) For $d = 10$, show that you can simulate observations of $W$ using the following program by graphing the histogram with the p.d.f. superimposed. Also graph the theoretical and empirical distribution functions.

```
m := 500;
for k to m do
if y < 0.5  then
XX[k] := evalf(-10/sqrt(2*99*y+1))
else
XX[k] := evalf(10/sqrt(99)/sqrt(1+100/99-2*y))
fi
od:
X := [seq(XX[k],k = 1 .. m)];
```

(c) Does your empirical data support the equality

$$P(|W-0| < k) = 1 - \frac{1}{k^2}?$$

In particular, for $k = 2, 3, 4, 5, 6$, is the proportion of observations that are within $k$ of $\mu = 0$ approximately equal to $1/k^2$?

(d) Does the distribution of $W$ satisfy the hypotheses of Chebyshev's inequality? In particular, find the mean and variance of $W$.

(e) For your simulated data, find the proportion of observations that are within $k$ (for $k = 2, 3, 4$) standard deviations ($\sqrt{v}$) of $\bar{x}$. How do these proportions compare with Chebyshev's guarantee?

**6.8–4** When taking successive Bernoulli trials with probability $p$ of success on each trial, let $Y_n$ equal the number of successes in $n$ trials. Then $\widehat{p} = Y_n/n$ is a point estimate of $p$. Chebyshev's inequality illustrates that $P(|Y_n/n - p| < \epsilon)$ converges to 1 as $n$ increases without bound, or that $Y_n/n = \widehat{p}$ converges to $p$ in probability. Illustrate this empirically by taking $n = 500$ Bernoulli observations with $p = 0.25$ and then use `PlotRunningAverage`.

# Chapter 7

# Tests of Statistical Hypotheses

## 7.1 Tests About Proportions

A statistical hypothesis is an assertion about a distribution of one or more random variables. A test of a statistical hypothesis $H_0$ is a procedure, based on the observed values of the items of a random sample, that leads to the acceptance or rejection of the hypothesis $H_0$.

Let $X$ have a distribution that depends on a parameter $\theta$. We shall consider a test of the simple (null) hypothesis $H_0: \theta = \theta_0$ against a composite alternative hypothesis $H_1: \theta \neq \theta_1$. Let $X_1, X_2, \ldots, X_n$ be a random sample from this distribution. Let $W = u(X_1, X_2, \ldots, X_n)$ denote the test statistic. The critical region, $C$, is the set of values of $W$ for which the hypothesis $H_0$ is rejected. Type I error occurs if $H_0$ is rejected when it is true. Type II error occurs if $H_0$ is accepted when it is false (that is, when $H_1$ is true). The significance level of the test is the probability of type I error, denoted by $\alpha$. We let $\beta$ denote the probability of type II error.

The $p$-value is the probability, under $H_0$, of all values of the test statistic that are as extreme (in the direction of rejection of $H_0$) as the observed value of the test statistic.

### EXERCISES

**Purpose:** The exercises illustrate tests of hypotheses, significance level, probability of type II error, and $p$-value of a test.

**7.1–1** Let $X_1, X_2, \ldots, X_n$ be a sequence of Bernoulli trials with probability of success, $p$, on each trial. We shall consider a test of the hypothesis $H_0: p = 0.25$ against the alternative hypothesis $H_1: p \neq 0.25$, using a significance level of $\alpha = 0.10$.

(a) If

$$Y = \sum_{i=1}^{n} X_i,$$

then $\widehat{p} = Y/n$ is approximately $N(0.25, (0.25)(0.75)/n)$ when $H_0$ is true. Two possible test statistics are

$$Z = \frac{Y/n - 0.25}{\sqrt{(0.25)(0.75)/n}} = \frac{\widehat{p} - 0.25}{\sqrt{(0.25)(0.75)/n}}$$

and

$$Z^* = \frac{Y/n - 0.25}{\sqrt{\widehat{p}(1-\widehat{p})/n}} = \frac{\widehat{p} - 0.25}{\sqrt{\widehat{p}(1-\widehat{p})/n}}$$

each of which is approximately $N(0,1)$ when $H_0$ is true. Find a constant $z_{\alpha/2}$ such that $P(Z \geq z_{\alpha/2}) = \alpha/2 = 0.05$. The hypothesis $H_0$ is rejected if $z \leq -z_{\alpha/2}$ or $z \geq z_{\alpha/2}$. If the $z^*$ statistic is used, $H_0$ is rejected if $z^* \leq -z_{\alpha/2}$ or $z^* \geq z_{\alpha/2}$.

(b) Generate 100 samples of $n = 40$ Bernoulli trials with $p = 0.25$. For each sample calculate $z$ and $z^*$.

```
A := [seq(BernoulliS(0.25,40),i = 1 .. 100)]:
Y := [seq(sum(A[i][j],j = 1 .. 40),i = 1 .. 100)]:
phat := [seq(Y[i]/40,i = 1 .. 100)]:
Z := [seq((phat[i]-0.25)/sqrt(0.25*0.75/40),
i = 1 .. 100)]:
Zstar := [seq(evalf((phat[i]-0.25)/
sqrt(phat[i]*(1-phat[i])/40)),i = 1 .. 100)]:
```

(c) Did $z$ and $z^*$ always lead to rejection of $H_0$ for the same random samples?

(d) Count the number of times that $H_0$ was rejected. Was $H_0$ rejected about 10% of the time both for $z$ and for $z^*$? You can use `sort` and inspect the lists `Z` and `Zstar`. Another way of arriving at the answers is given below.

```
z := NormalP(0,1,0.95);
for k from 1 to 100 do
Y := BernoulliS(0.25,40);
phat := Mean(Y);
Z := evalf((phat-0.25)/sqrt(0.25*0.75/40));
Zstar := evalf((phat-0.25 )/sqrt(phat*(1-phat)/40));
if z <= abs(Z) then rej[k] := 1
else rej[k] := 0
fi;
if z <= abs(Zstar) then rejstar[k] := 1
else rejstar[k] := 0
fi
od:
```

```
reject := convert(rej,list);
rejectstar := convert(rejstar,list);
Freq(reject,0 .. 1);
Locate(reject,1);
Freq(rejectstar,0 .. 1);
Locate(rejectstar,1);
```

**7.1–2** Let $X$ have a Bernoulli distribution, $b(1, p)$. We shall consider a test of $H_0: p = 0.50$ against $H_1: p = 0.25$. Let $X_1, X_2, \ldots, X_{10}$ be a random sample of size 10. Then

$$Y = \sum_{i=1}^{10} X_i$$

is $b(10, p)$. Define the critical region as $C = \{y: y/10 \le 0.35\} = \{y: y \le 3.5\}$.

(a) Find the value of $\alpha$. (Do not use a normal approximation.)

(b) Find the value of $\beta$. (Do not use a normal approximation.)

(c) Illustrate your answer to part (a) empirically by simulating 200 observations of $Y$ when $p = 0.50$. You may use `Y := BinomialS(10,0.50,200):`. Find the proportion of time that $H_0$ was rejected. Is this value close to $\alpha$?

(d) Illustrate your answer to part (b) empirically by simulating 200 observations of $Y$ when $p = 0.25$. You may use `Y := BinomialS(10,0.25,200):`. Find the proportion of time that $H_0$ was not rejected. Is this value close to $\beta$?

## Questions and Comments

**7.1–1** In Exercise 7.1-1, would you recommend that $z$ or that $z^*$ be used? Why?

**7.1–2** Note that the cumulative distribution function procedures (e.g., `NormalCDF`) can be used to find the $p$-value for most of the tests that we study.

# 7.2 Power Function and Sample Size

Let $X$ have a distribution that depends on a parameter $\theta$. We shall consider tests of a simple hypothesis $H_0 : \theta = \theta_0$ against a composite alternative hypothesis. Let $X_1, X_2, \ldots, X_n$ be a random sample from this distribution. Let $u(X_1, X_2, \ldots, X_n)$ be the test statistic and let $C$ be the critical region. The power function for this test is defined by

$$K(\theta) = P[u(X_1, X_2, \ldots, X_n) \in C;\ \theta].$$

That is, $K(\theta)$ is the probability of rejecting $H_0$ as a function of $\theta$.

## EXERCISES

**Purpose:** The exercises illustrate the power function for several different tests, both theoretically and empirically.

**7.2–1** Let $X$ have a Bernoulli distribution, $b(1, p)$. We shall test the hypothesis $H_0$: $p = 0.35$ against the alternative hypothesis $H_1 : p \neq 0.35$. Let $X_1, X_2, \ldots, X_{10}$ be a random sample of size 10 from this distribution and let

$$Y = \sum_{i=1}^{10} X_i.$$

Define the critical region as $C = \{y : |y - 3.5| > 2\} = \{y : y = 0, 1, 6, 7, 8, 9, 10\}$.

(a) Define the power function for this test using `BinomialCDF`.

(b) Graph the power function for $0 \le p \le 1$.

```
K := 1-BinomialCDF(10,p,5)+BinomialCDF(10,p,1);
P1 := plot(K,p = 0 .. 1,y = 0 .. 1):
plot(P1);
```

(c) Find the significance level of this test.

(d) For each of $p_1 = 0.05$, $p_2 = 0.15$, $\ldots, p_{10} = 0.95$, simulate 100 random samples of 10 Bernoulli trials with probability of success $p_i$. For each $p_i$, (i) obtain the test statistic, $y_i$, (ii) count the number of times that $H_0$ was rejected, (iii) calculate $r_i$, the proportion of times that $H_0$ was rejected.

```
m := 100; n := 10;
for i from 1 to 10 do
p[i] := 0.05+0.10*(i-1):
r[i] := 0:
Y := BinomialS(n,p[i],m);
for j from 1 to m do
if 2 < abs(Y[j]-3.5) then r[i] := r[i]+1
fi
od:
lprint('The number of rejections at ',p[i],
' is',r[i]);
od:
```

(e*) There are two ways to graph the empirical power function in order to compare it with the theoretical power function. (i) At each $p_i$, draw a vertical line segment from $[p_i, 0]$ to $[p_i, r_i]$. (ii) Graph the list $L := [p_1, r_1, p_2, r_2, \ldots, p_{10}, r_{10}]$. Add either or both of these graphs to your graph in part (b) of the theoretical power function.

```
S := {}:
for i from 1 to 10 do
p[i] := 0.05+0.10*(i-1):
S := S union {[p[i],0,p[i],r[i]/m]}
od:
P2 := plot(S):
plot({P1,P2});
Digits := 2:
L := [seq([p[i],evalf(r[i]/m)],i = 1 .. 10)];
P3 := plot(L):
plot({P1,P3});
plot({P1,P2,P3});
Digits := 10:
```

(f) At each $p_i$ let $W_i$ equal the number of times out of 100 that $H_0$ was rejected. Explain why $W_i$ has a binomial distribution and give the values of its parameters.

**7.2–2** Repeat Exercise 7.2-1 basing the test on a sample of size $n = 20$. Let the critical region be defined by $C = \{y: |y - 7| > 3.5\}$.

**7.2–3** Let $X$ have a normal distribution, $N(\mu, 144)$. We shall test the null hypothesis $H_0 : \mu = 50$ against the alternative hypothesis $H_1 : \mu > 50$. Let $X_1, X_2, \ldots, X_{16}$ be a random sample of size $n = 16$ from this distribution and let the critical region be defined by

$$C = \{\overline{x}: \overline{x} \geq 50 + (3)(1.282) = 53.846\}.$$

(a) Define the power function for this test using `NormalCDF`.

(b) Find the significance level of this test.

(c) Graph the power function for $50 \leq \mu \leq 60$.

(d) For each of $\mu_1 = 50,\ \mu_2 = 51, \ldots, \mu_{10} = 59$, simulate 50 random samples of size 16 from the normal distribution, $N(\mu_i, 144)$. For each $\mu_i$, (i) calculate the value of the test statistic, $\overline{x}$, (ii) count the number of times that $H_0$ was rejected, (iii) calculate $r_i$, the proportion of times that $H_0$ was rejected.

(e) There are two ways to graphically compare the empirical power function with the theoretical one. (i) At each $\mu_i$, draw a vertical line segment from $[\mu_i, 0]$ to $[\mu_i, r_i]$. (ii) Graph the list $L := [\mu_1, r_1, \mu_2, r_2, \ldots, \mu_{10}, r_{10}]$. Add either or both of these graphs to your graph in part (c) of the theoretical power function. (See Exercise 7.2-1.)

(f) At each $\mu_i$ let $W_i$ equal the number of times out of 50 that $H_0$ was rejected. Explain why $W_i$ has a binomial distribution and give the values of its parameters.

**7.2–4** Let $X$ have a normal distribution, $N(\mu, 144)$. We shall test the hypothesis $H_0\colon \mu = 50$ against the alternative hypothesis $H_1 : \mu > 50$ assuming that $\sigma^2$ is unknown. Given a sample of size $n = 16$, let the critical region be defined by $C = \{t : t \geq 1.341\}$ where $T = (\overline{X} - 50)/(S/4)$ has a $t$ distribution with $r = 15$ degrees of freedom when $H_0$ is true.

(a) Find the significance level of this test.

(b) For each of $\mu_1 = 50,\ \mu_2 = 51, \ldots, \mu_{10} = 59$, generate 50 random samples of size 16 from the normal distribution $N(\mu_i, 144)$. For each $\mu_i$, (i) obtain the value of the test statistic, $t$, (ii) count the number of times that $H_0$ was rejected, (iii) find $r_i$, the proportion of times that $H_0$ was rejected.

(c) Plot the empirically defined power function using the data that was generated in part (b). Either graph $L = [\mu_1, r_1, \mu_2, r_2, \ldots, \mu_{10}, r_{10}]$ or graph the vertical line segments from $[\mu_i, 0]$ to $[\mu_i, r_i], i = 1 \ldots 10$.

**7.2–5** Let $X$ have a Bernoulli distribution. We shall test $H_0\colon p \leq 0.40$ against the alternative hypothesis $H_1 : p > 0.40$. Given a random sample of size $n$, let $Y = \sum_{i=1}^{n} X_i$ be the test statistic and let the critical region be $C = \{y\colon y \geq c\}$.

(a) Let $n = 100$. On the same set of axes, graph the power functions corresponding to the critical regions $C_1 = \{y : y \geq 40\}, C_2 = \{y : y \geq 50\}$, and $C_3 = \{y\colon y \geq 60\}$.

(b) Let $C = \{y : y \geq 0.45n\}$. On the same set of axes, graph the power functions corresponding to the three sample sizes $n = 10$, $n = 100$, and $n = 200$.

**7.2–6** Let the distribution of $X$ be $N(\mu, 100)$. We shall test the null hypothesis $H_0\colon \mu = 80$ against the alternative hypothesis $H_1 : \mu < 80$. Given a random sample of size $n$, let the critical region be defined by $C = \{\overline{x}\colon \overline{x} \leq c\}$.

(a) Graph the three power functions corresponding to the three sample sizes $n = 16$, $n = 25$, and $n = 64$ on the same set of axes. Let $c = 77$. What are the respective significance levels?

(b) Graph the three power functions corresponding to the three values of $c = 77.436$, $c = 76.710$, and $c = 75.348$. Let $n = 25$. What are the respective significance levels?

## Questions and Comments

**7.2–1** In Exercises 7.2-1 and 7.2-2, what is the effect of increasing the sample size?

**7.2–2** How are the power functions in Exercises 7.2-3 and 7.2-4 related? What is the effect of not knowing the variance in Exercise 7.2-4?

**7.2–3** Based on 50 or 100 simulations at a particular value of the parameter being tested, e.g., $\theta$, how close should the empirical power function be to the theoretical power function? In particular let $W$ equal the number of samples out of 50 or 100 that lead to rejection of $H_0$. Is it clear that the distribution of W is $b(50, K(\theta))$ or $b(100, K(\theta))$?

## 7.3 Tests About One Mean and One Variance

Many of the exercises in the first two sections have already introduced tests about means and proportions. Recall that for tests about a mean, the test statistic is a function of the sample mean. Similarly, for tests about a variance, the test statistic is a function of the sample variance.

### EXERCISES

**Purpose:** The exercises illustrate empirically tests of hypotheses and power functions. A power function for a test about the mean of the exponential distribution is given. Practice in finding the $p$-value of a test is encouraged.

**7.3–1** Let $X_1, X_2, \ldots, X_n$ be a random sample of size $n$ from a normal distribution, $N(\mu, \sigma^2)$. We shall consider a test of the null hypothesis $H_0\colon \mu = 80$ against the two-sided alternative hypothesis $H_1 : \mu \neq 80$ at a significance level of $\alpha = 0.10$ when $n = 8$.

(a) If it is known that $\sigma^2 = 25$, then $Z = (\overline{X} - 80)/(5/\sqrt{n})$ is N(0,1) when $H_0$ is true. Find a constant $z_{\alpha/2}$ such that $P(Z \geq z_{\alpha/2}) = \alpha/2 = 0.05$. The hypothesis $H_0$ is rejected if $z \leq -z_{\alpha/2}$ or if $z \geq z_{\alpha/2}$. Equivalently $H_0$ is rejected if $\overline{x} \leq 80 - z_{\alpha/2}(5/\sqrt{n})$ or $\overline{x} \geq 80 + z_{\alpha/2}(5/\sqrt{n})$.

(b) If $\sigma^2$ is unknown, we can use the statistic $T = (\overline{X} - 80)/(S/\sqrt{n})$ which has a $t$ distribution with $r = n - 1$ degrees of freedom when $H_0$ is true. Find a constant $t_{\alpha/2}(n-1)$ such that $P(T \geq t_{\alpha/2}(n-1)) = \alpha/2 = 0.05$ when $n = 8$. The hypothesis $H_0$ is rejected if $t \leq -t_{\alpha/2}(n-1)$ or $t \geq t_{\alpha/2}(n-1)$ or, equivalently, when $\overline{x} \leq 80 - t_{\alpha/2}(n-1)(s/\sqrt{n})$ or $\overline{x} \geq 80 + t_{\alpha/2}(n-1)(s/\sqrt{n})$.

(c) Generate 100 samples of size $n = 8$ from the normal distribution $N(80, 25)$. For each sample, calculate and print $z$, the $p$-value of $z$, $t$, and the $p$-value of $t$.

(d) Count the number of times that $H_0$ was rejected when it was assumed that $\sigma^2 = 25$ was known and the $z$ statistic was used. Was $H_0$ rejected about 10% of the time?

(e) Count the number of times that $H_0$ was rejected when the $t$ statistic was used. Was $H_0$ rejected about 10% of the time?

(f) Did the $z$ and $t$ statistics always lead to rejection of $H_0$ for the same samples? Explain.

**7.3–2** Let $X$ have a normal distribution, $N(\mu, 64)$. We shall consider a test of the hypothesis $H_0\colon \mu = 75$ against the alternative hypothesis $H_1\colon \mu = 80$. Given a random sample of size 16, $X_1, X_2, \ldots, X_{16}$, let the critical region be defined by

$$C = \{\overline{x}\colon \overline{x} \geq 78\} = \{\overline{x}\colon (\overline{x} - 75)/(8/4) \geq 1.5\}.$$

(a) Find the value of $\alpha$.

(b) Find the value of $\beta$.

(c) Illustrate your answer to part (a) empirically. In particular, generate 50 samples of size 16 from the normal distribution, $N(75, 64)$. What proportion of these samples led to rejection of $H_0$?

(d) Illustrate your answer to part (b) empirically. In particular, generate 50 samples of size 16 from the normal distribution, $N(80, 64)$. What proportion of these samples led to acceptance of $H_0$?

**7.3–3** In Exercise 7.3-2 assume that $\sigma^2$ is unknown so that the test is defined in terms of a $t$ statistic. Let

$$C = \left\{\overline{x}\colon \frac{\overline{x} - 75}{s/4} \geq 1.75\right\}.$$

Work Parts c and d of Exercise 7.3-2 under these assumptions.

**7.3–4** Let X have a normal distribution, $N(\mu, \sigma^2)$. We shall consider a test of $H_0\colon \sigma^2 = 50$ against $H_1\colon \sigma^2 = 100$. Let $X_1, X_2, \ldots, X_{20}$ be a random sample of size 20. Let the critical region be defined by $C = \{s^2\colon s^2 \geq 70\}$.

(a) How is $(n-1)S^2/\sigma^2$ distributed?

(b) Find the value of $\alpha$.

(c) Find the value of $\beta$.

(d) Illustrate your answer to part (b) empirically. In particular, generate 50 samples of size 20 from a normal distribution, $N(6, 50)$. Print the values of the 50 $s^2$'s. What proportion of these samples led to the rejection of $H_0$? Is this number close to $\alpha$?

(e) Illustrate your answer to part (c) empirically. In particular generate 50 samples of size 20 from a normal distribution, $N(6, 100)$. Print the values of the 50 $s^2$'s. What proportion of these samples led to the acceptance of $H_0$? Is this number close to $\beta$?

**7.3–5** Let $X_1, X_2, \ldots, X_{10}$ be a random sample of size 10 from the normal distribution $N(\mu, \sigma^2)$. We shall test the null hypothesis $H_0\colon \sigma^2 = 47.5$ against the alternative hypothesis $H_1 : \sigma^2 < 47.5$.

(a) Find a constant $c$ so that $C = \{s^2 : s^2 \le c\}$ is a critical region of size $\alpha = 0.05$. **Hint**: The statistic $(10-1)S^2/47.5$ is $\chi^2(9)$ when $H_0$ is true.

(b) Define the power function for this test.

(c) For $\sigma^2 = 2.5, 7.5, 12.5, \ldots, 47.5$, generate 50 random samples of size 10 from the normal distribution, $N(0, \sigma^2)$. For each of the 10 values of $\sigma^2$ count the number of times that $H_0$ was rejected.

(d) Plot the power function defined in part (b) along with the power function defined empirically by the data generated in part (c).

**7.3–6** Let $X_1, X_2, \ldots, X_{12}$ be a random sample of size 12 from an exponential distribution with mean $\theta$. That is, the p.d.f. is

$$f(x;\theta) = (1/\theta)e^{-x/\theta}, \quad 0 < x < \infty.$$

We shall test the null hypothesis $H_0\colon \theta = 9$ against the two-sided alternative hypothesis $H_1 : \theta \ne 9$. Recall that

$$\frac{2}{\theta}\sum_{i=1}^{12} X_i \text{ is } \chi^2(24).$$

(a) Find constants $a$ and $b$ so that

$$P\left[\frac{2}{9}\sum_{i=1}^{12} X_i \le a\right] = 0.05$$

and

$$P\left[\frac{2}{9}\sum_{i=1}^{12} X_i \ge b\right] = 0.05.$$

Then a critical region of size $\alpha = 0.10$ is

$$C = \left\{(x_1, \ldots, x_{12})\colon \sum_{i=1}^{12} x_i \le \frac{9a}{2} \text{ or } \sum_{i=1}^{12} x_i \ge \frac{9b}{2}\right\}.$$

(b) Define the power function for this test.

(c) For $\theta = 1, 3, 5, \ldots, 19$, generate 50 random samples of size 12 from an exponential distribution with a mean of $\theta$. For each $\theta$ count the number of times that $H_0$ was rejected.

(d) Plot the power function defined in part (b) along with the empirically defined power function using the data generated in part (c).

**7.3–7** Find the $p$-value for some of the exercises in your textbook. For each of them, graph the appropriate p.d.f. (normal, $t$, $\chi^2$, $F$) and show the $p$-value on your graph.

## Questions and Comments

**7.3–1** Did your theoretical and empirical results agree? Combine your empirical results with those of other class members.

**7.3–2** The distribution function procedures can be used to find $p$-values.

# 7.4 Tests of the Equality of Two Variances

For tests comparing two variances, the test statistic is a function of the ratio of the sample variances.

## EXERCISES

**Purpose:** The exercises illustrate tests of hypotheses about two variances, including some power function illustrations.

**7.4–1** Let $X_1, X_2, \ldots, X_9$ and $Y_1, Y_2, \ldots, Y_6$ be random samples of sizes $n = 9$ and $m = 6$ from independent normal distributions, $N(\mu_X, \sigma_X^2)$ and $N(\mu_Y, \sigma_Y^2)$, respectively. Consider a test of the hypothesis $H_0 : \sigma_X^2/\sigma_Y^2 = 1$ against the alternative hypothesis $H_1 : \sigma_X^2/\sigma_Y^2 = 5$. Let the critical region be defined by

$$C = \{(s_X^2, s_Y^2) : s_X^2/s_Y^2 \geq c\}.$$

(a) Find the value of $c$ so that $\alpha = 0.10$. **Hint:**

$$F = \frac{S_X^2/\sigma_X^2}{S_Y^2/\sigma_Y^2} = \frac{S_X^2}{S_Y^2}$$

has an $F$ distribution with $r_1 = 8$ and $r_2 = 5$ degrees of freedom when $H_0$ is true.

(b) Find the value of $\beta$. **Hint:**

$$F = \frac{S_X^2/\sigma_X^2}{S_Y^2/\sigma_Y^2} = \frac{S_X^2}{5S_Y^2}$$

has an $F$ distribution with $r_1 = 8$ and $r_2 = 5$ degrees of freedom when $H_1$ is true.

(c) Illustrate your answer to part (a) empirically. In particular, generate 50 samples of sizes 9 and 6 from the standard normal distribution and calculate the values of $s_X^2/s_Y^2$. What proportion of these samples led to rejection of $H_0$? Is this number close to $\alpha$?

(d) Illustrate your answer to part (b) empirically. In particular, generate 50 samples of sizes 9 and 6 from the normal distributions, $N(0,5)$ and $N(0,1)$, respectively. Calculate and print the values of $s_X^2/s_Y^2$. What proportion of these samples led to acceptance of $H_0$? Is this number close to $\beta$?

**7.4–2** Let $X_1, X_2, \ldots, X_9$ and $Y_1, Y_2, \ldots, Y_6$ be random samples of sizes $n = 9$ and $m = 6$ from independent normal distributions, $N(\mu_X, \sigma_X^2)$ and $N(\mu_Y, \sigma_Y^2)$, respectively. Consider a test of the hypothesis $H_0: \sigma_X^2/\sigma_Y^2 = 1$ against the alternative hypothesis $H_1: \sigma_X^2/\sigma_Y^2 > 1$.

(a) Find the value of $c$ so that

$$C = \{(x_1, \ldots, x_9, y_1, \ldots, y_6) : s_X^2/s_Y^2 \geq c\}$$

is a critical region of size $\alpha = 0.10$.

(b) Let $\theta = \sigma_X^2/\sigma_Y^2$. Define the power function for this test as a function of $\theta$.

(c) Plot the power function.

**7.4–3** Repeat Problem 7.4–2 for $\alpha = 0.05$.

**7.4–4** Repeat Problem 7.4–2 for $\alpha = 0.025$.

**7.4–5** Repeat Problem 7.4–2 for $\alpha = 0.01$.

**7.4–6** Let $X_1, X_2, \ldots, X_9$ and $Y_1, Y_2, \ldots, Y_6$ be random samples of sizes $n = 9$ and $m = 6$ from independent normal distributions, $N(\mu_X, \sigma_X^2)$ and $N(\mu_Y, \sigma_Y^2)$, respectively. We shall test the null hypothesis $H_0: \sigma_X^2/\sigma_Y^2 = 1$ against the alternative hypothesis $H_1: \sigma_X^2/\sigma_Y^2 > 1$.

(a) Find the value of $c$ so that

$$C = \{(x_1, \ldots, x_9, y_1, \ldots, y_6) : s_X^2/s_Y^2 \geq c\}$$

is a critical region of size $\alpha = 0.10$.

(b) Let $\theta = \sigma_X^2/\sigma_Y^2$. Define the power function for this test as a function of $\theta$.

(c) Let $\sigma_Y^2 = 1$, let $\sigma_X^2 = 1, 2, \ldots, 10$. Thus $\theta = 1, 2, \ldots, 10$. For each of the 10 values of $\theta$ simulate 50 random samples of sizes $n = 9$ and $m = 6$ from the normal distributions, $N(0, \sigma_X^2)$ and $N(0,1)$, respectively, and count the number of times that $H_0: \theta = 1$ was rejected.

(d) Plot the power function defined in part (b) along with the power function defined empirically by the data generated in part (c).

## Questions and Comments

**7.4–1** What is the effect of the value of the significance level on the power function? See, for example, Exercises 7.4–2, 7.4–3, 7.4–4, and 7.4–5.

# 7.5 Graphical Methods

Given random samples from two independent distributions, a quantile–quantile ($q$-$q$) plot is a scatter diagram of the points $(x_i, y_i)$, where $x_i$ and $y_i$ are the same quantiles of each of the two random samples.

Given a random sample, a scatter diagram of the quantiles of the sample with the quantiles of a theoretical distribution provides a $q$–$q$ plot that can be used for model fitting.

## EXERCISES

**Purpose:** The exercises illustrate $q$–$q$ plots using random samples from similar and from different distributions so you can see the effect on the $q$–$q$ plot.

**7.5–1** Use `QQ` to construct quantile–quantile plots, taking random samples of size 25 from each of the following pairs of distributions. Interpret the $q$–$q$ plots.

(a) $U(0, 10)$ and $U(0, 10)$,

(b) $U(10, 20)$ and $U(0, 10)$,

(c) $U(0, 20)$ and $U(5, 15)$,

(d) $U(5, 15)$ and $U(0, 20)$,

(e) The distribution with p.d.f. $f(x) = 3x^2$, $0 < x < 1$, and the distribution with p.d.f. $f(x) = (1/3)x^{-2/3}$, $0 < x < 1$,

(f) The converse of part (e),

(g) $N(1, 1)$ and the exponential with mean $\theta = 1$.

**7.5–2** Using a random sample of size $n = 19$ from the standard normal distribution, make a quantile–quantile plot using the ordered observations as the $x$–coordinates and the respective quantiles of the $N(0, 1)$ distribution as the $y$–coordinates. Note that these quantiles, for $k = 1, 2, \ldots, 19$, can be found using `NormalP(0,1,k/20);`.

**7.5–3** Using a random sample of size $n = 19$ from $N(50, 16)$, make a quantile-quantile plot using the ordered observations as the $x$–coordinates and the respective quantiles of either the $N(0, 1)$ distribution or the $N(50, 16)$ distribution as the $y$–coordinates.

**7.5–4** Using a random sample of size $n = 19$ from the exponential distribution with mean $\theta = 10$, make a quantile–quantile plot using the ordered observations as the $x$–coordinates and the respective quantiles of this exponential distribution as the $y$–coordinates. You may use `ExponentialP(10,k/20);` to find these quantiles.

## 7.6 Likelihood Ratio Tests

Assume that the p.d.f. of $X$ is $f(x;\theta)$, $\theta \in \Omega$. Let $\omega \subset \Omega$. Consider a test of the hypothesis

$$H_0\colon \theta \in \omega$$

against

$$H_1\colon \theta \in \omega' = \Omega - \omega.$$

The likelihood ratio is the quotient

$$\lambda = \frac{L(\widehat{\omega})}{L(\widehat{\Omega})}$$

where $L(\widehat{\omega})$ is the maximum of the likelihood function with respect to $\theta$ when $\theta \in \omega$ and $L(\widehat{\Omega})$ is the maximum likelihood function with respect to $\theta$ when $\theta \in \Omega$.

To test $H_0\colon \theta \in \omega$ against $H_1\colon \theta \in \omega'$, the critical region for the likelihood ratio test is the set of points in the sample space for which

$$\lambda = \frac{L(\widehat{\omega})}{L(\widehat{\Omega})} \le k$$

where $0 < k < 1$ and $k$ is selected so that the test has a desired significance level $\alpha$.

## EXERCISES

**Purpose:** The exercises illustrate the likelihood ratio test criterion for testing a mean and testing a variance for a normally distributed random variable.

**7.6–1** Let $X$ be $N(\mu, 5)$.

(a) Show that a critical region given by the likelihood ratio test criterion for testing $H_0\colon \mu = 162$ against $H_1\colon \mu \neq 162$ is given by

$$\lambda = G(\overline{x}) = \exp[(-n/10)(\overline{x} - 162)^2]$$

(b) Construct the graph of this likelihood ratio for $158 \le \overline{x} \le 166$.

**7.6–2** Let $X$ be $N(\mu, \sigma^2)$. We shall test $H_0\colon \sigma^2 = \sigma_0^2$ against $H_1 : \sigma^2 \neq \sigma_0^2$.

(a) Show that a critical region given by the likelihood ratio test criterion is given by

$$\lambda = g(w) = n^{(-n/2)} e^{n/2} w^{n/2} e^{-w/2} \leq k$$

or

$$h(w) = w^{n/2} e^{-w/2} \leq k n^{n/2} e^{-n/2} = c$$

where $w = \sum_1^n (x_i - \overline{x})^2/\sigma_0^2$ and $k$ or $c$ is selected to yield the desired significance level.

(b) How is $W = \sum_1^n (x_i - \overline{x})^2/\sigma_0^2$ distributed when $H_0$ is true?

(c) Construct the graph of $\lambda = g(w)$ and construct the graph of $y = h(w)$ when $n = 10$ for $0 \leq w \leq 30$.

(d) The critical region is given by $C = \{w : w \leq a \quad \text{or} \quad w \geq b\}$ where $g(a) = g(b)$ and $P(a < W < b;\ H_0) = 1 - \alpha$, or $P(W \in C;\ H_0) = \alpha$. For $n = 10$ and $\alpha = 0.10$, solve for $a$ and $b$.

(e) What is the relation between your answers in part (d) and the chi-square values for a confidence interval of minimum length for $\sigma$?

# Chapter 8

# Linear Models

## 8.1 Tests of the Equality of Several Means

Consider $m$ mutually independent random variables that have normal distributions with unknown means $\mu_1, \mu_2, \ldots, \mu_m$, respectively, and unknown but common variance, $\sigma^2$. Let $X_{i1}, X_{i2}, \ldots, X_{in_i}$ represent a random sample of size $n_i$ from the normal distribution $N(\mu_i, \sigma^2)$, $i = 1, 2, \ldots, m$. Let $n = n_1 + n_2 + \cdots + n_m$ and

$$\begin{aligned} SS(TO) &= \sum_{i=1}^{m}\sum_{j=1}^{n_i}(X_{ij} - \overline{X}_{\cdot\cdot})^2 \\ &= \sum_{i=1}^{m}\sum_{j=1}^{n_i}(X_{ij} - \overline{X}_{i\cdot})^2 + \sum_{i=1}^{k} n_i(\overline{X}_{i\cdot} - \overline{X}_{\cdot\cdot})^2 \\ &= SS(E) + SS(T) \end{aligned}$$

where $\overline{X}_{\cdot\cdot} = (1/n)\sum_{i=1}^{m}\sum_{j=1}^{n_i} X_{ij}$ and $\overline{X}_{i\cdot} = (1/n_i)\sum_{j=1}^{n_i} X_{ij}$. A critical region for testing the hypothesis $H_0: \mu_1 = \mu_2 = \cdots = \mu_m = \mu$ against all alternatives can be defined in terms of the test statistic

$$F = \frac{SS(T)/[\sigma^2(m-1)]}{SS(E)/[\sigma^2(n-m)]} = \frac{SS(T)/(m-1)}{SS(E)/(n-m)}$$

which has an $F$ distribution with $r_1 = m - 1$ and $r_2 = n - m$ degrees of freedom when $H_0$ is true. The hypothesis is rejected if $F \geq F_\alpha(m-1, n-m)$.

### EXERCISES

**Purpose:** The exercises give applications of this test.

**8.1–1** Three groups of 12 albino rats were run in three different field conditions, one group per field, for a physiological psychology experiment. The first was an open field. In the second, two mechanical toys were placed in the field with the rat — on the assumption that the noise and the movement of the toy animals would be a fearful stimulus to the rat. In the third, a bright light was projected over the field with a carousel projector, with the light on for five-second intervals, alternated with 0.5 second light off intervals. The variable measured was the number of squares traveled over a ten-minute period.

| | | | | | | |
|---|---|---|---|---|---|---|
| Field 1 ($X_1$): | 21 | 3 | 34 | 152 | 46 | 37 |
| | 108 | 200 | 117 | 111 | 12 | 102 |
| Field 2 ($X_2$): | 86 | 25 | 57 | 57 | 146 | 28 |
| | 118 | 114 | 115 | 76 | 90 | 79 |
| Field 3 ($X_3$): | 71 | 18 | 35 | 202 | 96 | 94 |
| | 189 | 220 | 190 | 200 | 74 | 80 |

Let $\mu_i$, $i = 1, 2, 3$, denote the mean number of squares traversed over a ten-minute period for each of the three fields, respectively. We are interested in testing the hypothesis $H_0\colon \mu_1 = \mu_2 = \mu_3 = \mu$.

(a) Use the subroutine `Anova1` to perform this test. Interpret the $p$–value and give a conclusion for the test. (These data are stored in `STAT.DAT` as `D811`.)

(b) Draw box-and-whisker diagrams on the same figure. **Hint**: If the data is in a list of lists called `D811`, you may use `BoxWhisker(D811[1..3]);`.

**8.1–2** An instructor gave a quiz to three different groups of students and observed the following scores.

| | | | | | | | |
|---|---|---|---|---|---|---|---|
| $X_1$ : | 13 | 10 | 12 | 7 | 8 | 9 | 11 |
| $X_2$ : | 13 | 16 | 13 | 14 | 15 | 12 | 10 |
| $X_3$ : | 15 | 12 | 14 | 10 | 10 | 11 | 9 |

(a) Let $\mu_i$, $i = 1, 2, 3$, denote the mean score for group $i$. Use `Anova1` to test $H_0\colon \mu_1 = \mu_2 = \mu_3 = \mu$.

(b) Draw box-and-whisker diagrams on the same figure.

**8.1–3** A percent composition was done on rock suites from Hawaii, Boulder, and the Michigan upper peninsula. The observed percentages of $TiO_2$ within 4 random samples from each location were as follows:

| | | | | |
|---|---|---|---|---|
| Boulder: | 0.52 | 0.18 | 0.56 | 0.59 |
| Hawaii: | 1.97 | 0.37 | 2.90 | 0.59 |
| U.P.: | 1.71 | 0.06 | 1.90 | 0.43 |

Let $\mu_i$ denote the mean percentages. Use `Anova1` to test $H_0\colon \mu_1 = \mu_2 = \mu_3$.

**8.1–4** To test whether three varieties of corn give equal yield, each variety was planted on each of four different test sites yielding the following numbers of bushels per acre.

| | | | | |
|---|---|---|---|---|
| Variety 1: | 129.6 | 121.9 | 122.8 | 147.2 |
| Variety 2: | 132.2 | 135.6 | 158.8 | 151.0 |
| Variety 3: | 150.1 | 133.5 | 146.3 | 143.4 |

Let $\mu_i$, $i = 1, 2, 3$, denote the yield for variety $i$. Test $H_0\colon \mu_1 = \mu_2 = \mu_3$.

**8.1–5** Use `Anova1` to work some exercises from your textbook. For each of them, when the sample sizes are seven or larger, draw multiple box-and-whisker diagrams and use these to interpret your conclusions.

## Questions and Comments

**8.1–1** Use `Anova1` to work the exercises in your textbook. Be sure to interpret the $p$–value of each test.

# 8.2 Two–Factor Analysis of Variance

Assume that there are two factors, one of which has $a$ levels (rows) and the other has $b$ levels (columns). Let $X_{ij}$ denote the observation in the $i$th row and $j$th column and let $X_{ij}$ be $N(\mu_{ij}, \sigma^2)$, $i = 1, \ldots, a$, and $j = 1, \ldots, b$. We further assume that $\mu_{ij} = \mu + \alpha_i + \beta_j$, where

$$\sum_{i=1}^{a} \alpha_i = 0 \quad \text{and} \quad \sum_{j=1}^{b} \beta_i = 0.$$

Letting $\overline{X}_{i\cdot} = (1/b)\sum_{j=1}^{b} X_{ij}$, $\overline{X}_{\cdot j} = (1/a)\sum_{i=1}^{a} X_{ij}$, and $\overline{X}_{\cdot\cdot} = (1/ab)\sum_{j=1}^{b}\sum_{i=1}^{a} X_{ij}$, we have

$$SS(TO) = \sum_{j=1}^{b}\sum_{i=1}^{a}(X_{ij} - \overline{X}_{\cdot\cdot})^2$$

$$
\begin{aligned}
&= b\sum_{i=1}^{a}(\overline{X}_{i\cdot} - \overline{X}_{\cdot\cdot})^2 + a\sum_{j=1}^{b}(\overline{X}_{ij} - \overline{X}_{\cdot\cdot})^2 \\
&\quad + \sum_{j=1}^{b}\sum_{i=1}^{a}(X_{ij} - \overline{X}_{i\cdot} - \overline{X}_{\cdot j} + \overline{X}_{\cdot\cdot})^2 \\
&= SS(A) + SS(B) + SS(E).
\end{aligned}
$$

A test of the hypothesis of $H_R\colon \alpha_1 = \alpha_2 = \cdots = \alpha_a = 0$, no row effect or no effect due to the levels of factor $A$, can be based on the statistic

$$F_R = \frac{SS(A)/(a-1)}{SS(E)/[(a-1)(b-1)]}$$

which has an $F$ distribution with $a-1$ and $(a-1)(b-1)$ degrees of freedom when $H_R$ is true. The hypothesis $H_R$ is rejected if $F_R \geq F_\alpha(a-1, [a-1][b-1])$.

Similarly a test of the hypothesis $H_C\colon \beta_1 = \beta_2 = \cdots = \beta_b = 0$, can be based on

$$F_C = \frac{SS(C)/(b-1)}{SS(E)/[(a-1)(b-1)]}$$

which has an $F$ distribution with $b-1$ and $(a-1)(b-1)$ degrees of freedom when $H_C$ is true. The hypothesis $H_C$ is rejected if $F_C > F_\alpha(b-1, [a-1][b-1])$.

Suppose that in a two-way classification problem, $c > 1$ independent observations are taken per cell. Assume that $X_{ijk}$, $i = 1, 2, \ldots, a$, $j = 1, 2, \ldots, b$, and $k = 1, 2, \ldots, c$ are $n = abc$ mutually independent random variables having normal distributions with a common, but unknown, variance $\sigma^2$. The mean of $X_{ijk}$, $k = 1, 2, \ldots, c$ is $\mu_{ij} = \mu + \alpha_i + \beta_j + \gamma_{ij}$ where

$$\sum_{i=1}^{a}\alpha_i = 0 \quad \text{and} \quad \sum_{j=1}^{b}\beta_j = 0,$$

$$\sum_{i=1}^{a}\gamma_{ij} = 0 \quad \text{and} \quad \sum_{j=1}^{b}\gamma_{ij} = 0.$$

The parameter $\gamma_{ij}$ is called the interaction associated with cell $(i, j)$.

Letting

$$\overline{X}_{ij\cdot} = (1/c)\sum_{k=1}^{c}X_{ijk}, \quad \overline{X}_{i\cdot\cdot} = (1/bc)\sum_{j=1}^{b}\sum_{k=1}^{c}X_{ijk}, \quad \overline{X}_{\cdot j\cdot} = (1/ac)\sum_{i=1}^{a}\sum_{k=1}^{c}X_{ijk}$$

and

$$\overline{X}_{\cdots} = (1/abc)\sum_{i=1}^{a}\sum_{j=1}^{b}\sum_{k=1}^{c}X_{ijk}$$

we have

$$\begin{aligned}
SS(TO) &= \sum_{i=1}^{a}\sum_{j=1}^{b}\sum_{k=1}^{c}(X_{ijk} - \overline{X}_{...})^2 \\
&= bc\sum_{i=1}^{a}(\overline{X}_{i..} - \overline{X}_{...})^2 + ac\sum_{j=1}^{b}(\overline{X}_{.j.} - \overline{X}_{...})^2 \\
&\quad + c\sum_{i=1}^{a}\sum_{j=1}^{b}(\overline{X}_{ij.} - \overline{X}_{i..} - \overline{X}_{.j.} + \overline{X}_{...})^2 + \sum_{i=1}^{a}\sum_{j=1}^{b}\sum_{k=1}^{c}(X_{ijk} - \overline{X}_{ij.})^2 \\
&= SS(A) + SS(B) + SS(AB) + SS(E).
\end{aligned}$$

A test of the hypothesis $H_I\colon \gamma_{ij} = 0,\ i = 1, 2, \ldots, a,\ j = 1, 2, \ldots, b$, can be based on the statistic

$$F_I = \frac{SS(AB)/[(a-1)(b-1)]}{SS(E)/[ab(c-1)]}$$

which has an $F$ distribution with $(a-1)(b-1)$ and $ab(c-1)$ degrees of freedom when $H_I$ is true.

A test of the hypothesis $H_R\colon \alpha_1 = \alpha_2 = \cdots = \alpha_a = 0$ can be based on the statistic

$$F_R = \frac{SS(A)/(a-1)}{SS(E)/[ab(c-1)]}$$

which has an $F$ distribution with $a-1$ and $ab(c-1)$ degrees of freedom when $H_R$ is true.

A test of the hypothesis $H_C\colon \beta_1 = \beta_2 = \cdots = \beta_b = 0$ can be based on the statistic

$$F_C = \frac{SS(B)/(b-1)}{SS(E)/[ab(c-1)]}$$

which has an $F$ distribution with $b-1$ and $ab(c-1)$ degrees of freedom when $H_C$ is true.

## EXERCISES

**Purpose:** The exercises illustrate how the computer can be used to solve analysis of variance problems.

**8.2–1** The National Assessment of Educational Progress (NAEP) published the following data in the report *Trends in Academic Progress*, November, 1991. The table gives the mathematics scores, classified by number of hours of TV watching each day and the age of the student.

| | **Age** | | |
|---|---|---|---|
| **TV watching time** | Age 9 | Age 13 | Age 17 |
| 0–2 hours | 311 | 276 | 230 |
| 3–5 hours | 299 | 270 | 233 |
| 6 or more hours | 285 | 256 | 220 |

Assume that each score is an observation of a $N(\mu_{ij}, \sigma^2)$ random variable where the mean $\mu_{ij} = \mu + \alpha_i + \beta_j$, $i = 1, 2, 3; j = 1, 2, 3$. Use **Anova2s** to test each of the following hypotheses. Give an interpretation of your conclusions using the $p$-values.

(a) $H_A\colon \alpha_1 = \alpha_2 = \alpha_3 = 0$ (no row effect),

(b) $H_B\colon \beta_1 = \beta_2 = \beta_3 = 0$ (no column effect).

**8.2–2** A student in psychology was interested in testing how water consumption by rats would be affected by a particular drug. She used two levels of one attribute, namely, drug and placebo, and four levels of a second attribute, namely, male ($M$), castrated ($C$), female ($F$), and ovariectomized ($O$). For each cell she observed five rats. The amount of water consumed in milliliters per 24 hours is listed in Table 8.2–1. Use **Anova2m** to test the hypotheses (your choice of significance level)

(a) $H_I\colon \gamma_{ij} = 0,\ i = 1, 2,\ j = 1, 2, 3, 4,$

(b) $H_R\colon \alpha_1 = \alpha_2 = 0,$

(c) $H_C\colon \beta_1 = \beta_2 = \beta_3 = \beta_4 = 0.$

(d) Draw eight box-and-whisker diagrams on the same graph and use them to help to interpret your ANOVA summary table.

| | M | C | F | O |
|---|---|---|---|---|
| | 30.4 | 29.6 | 37.2 | 27.2 |
| | 32.5 | 29.8 | 31.2 | 31.0 |
| Drug | 31.8 | 25.4 | 33.0 | 26.2 |
| | 34.6 | 36.2 | 33.2 | 28.4 |
| | 34.4 | 28.2 | 31.6 | 29.2 |
| | 36.6 | 35.8 | 24.4 | 31.4 |
| | 35.0 | 37.8 | 27.0 | 32.0 |
| Placebo | 34.0 | 43.2 | 24.4 | 29.4 |
| | 35.6 | 42.0 | 27.4 | 32.4 |
| | 33.6 | 41.0 | 30.4 | 30.8 |

**Table 8.2–1**

**8.2–3** The student in Exercise 8.2–2 was also interested in testing how the activity of rats would be affected by this drug. Table 8.2–2 lists the number of revolutions of an activity wheel in 24 hours. Test the hypotheses and draw box-and-whisker diagrams as requested in Exercise 8.2–1 using these data.

| | M | C | F | O |
|---|---|---|---|---|
| Drug | 1401 | 209 | 2029 | 409 |
| | 617 | 451 | 8370 | 487 |
| | 226 | 215 | 5372 | 412 |
| | 289 | 483 | 2738 | 561 |
| | 440 | 270 | 8725 | 659 |
| Placebo | 164 | 837 | 2892 | 514 |
| | 306 | 705 | 3728 | 450 |
| | 170 | 577 | 3131 | 854 |
| | 398 | 724 | 1655 | 683 |
| | 263 | 500 | 2039 | 656 |

**Table 8.2–2**

## Questions and Comments

**8.2–1** The assumptions of normality and homogeneous variances can be relaxed in applications with little change in the significance levels of the resulting tests. Use empirical methods to determine how relaxed these assumptions are.

# 8.3 Regression Analysis

When looking at relations between two variables, sometimes one of the variables, say $x$, is known in advance and there is interest in predicting the value of a random variable $Y$. To learn something about a particular situation, $n$ observations are made resulting in the $n$ pairs of data, $(x_1, y_1), (x_2, y_2), \ldots, (x_n, y_n)$. We would then like to estimate $E(Y) = \mu(x)$, the mean value of $Y$ for a given $x$. Sometimes $\mu(x)$ is of a given form such as linear, quadratic, or exponential. That is, it could be true that $E(Y) = \alpha + \beta x$ or that $E(Y) = \alpha + \beta x + \gamma x^2$, or $E(Y) = \alpha e^{\beta x}$. Given the set of data, our goal is to find good estimates of the unknown parameters, $\alpha, \beta$, and possibly $\gamma$.

To begin with, we assume that

$$Y_i = \alpha + \beta(x_i - \overline{x}) + \epsilon_i,$$

where $\epsilon_i$, for $i = 1, 2, \ldots, n$, are independent $N(0, \sigma^2)$ random variables.

Maximum likelihood estimates of $\alpha, \beta$, and $\sigma^2$ are

$$\widehat{\alpha} = \overline{Y},$$

$$\widehat{\beta} = \frac{\sum_{i=1}^{n} Y_i(x_i - \overline{x})}{\sum_{i=1}^{n}(x_i - \overline{x})^2},$$

and

$$\widehat{\sigma^2} = \frac{1}{n}\sum_{i=1}^{n}[Y_i - \widehat{\alpha} - \widehat{\beta}(x_i - \overline{x})]^2.$$

Let

$$\widehat{Y_i} = \widehat{\alpha} + \widehat{\beta}(x_i - \overline{x}),$$

the predicted mean value of $Y_i$. The difference

$$Y_i - \widehat{Y_i} = Y_i - \widehat{\alpha} - \widehat{\beta}(x_i - \overline{x})$$

is called the $i$th residual, $i = 1, 2, \ldots, n$. A scatterplot of the values of $x$ with their respective residuals is often helpful in seeing whether linear regression is an appropriate model.

It is possible to form confidence intervals for $\alpha, \beta$, and $\sigma^2$ using appropriate $t$ statistics. The endpoints for the respective $100(1 - \alpha)\%$ confidence intervals are as follows:

$$\widehat{\beta} \pm t_{\alpha/2}(n-2)\sqrt{\frac{n\widehat{\sigma^2}}{(n-2)\sum_{i=1}^{n}(x_i - \overline{x})^2}},$$

$$\widehat{\alpha} \pm t_{\alpha/2}(n-2)\sqrt{\frac{\widehat{\sigma^2}}{n-2}},$$

and

$$\left[\frac{n\widehat{\sigma^2}}{\chi^2_{\alpha/2}(n-2)}, \frac{n\widehat{\sigma^2}}{\chi^2_{1-\alpha/2}(n-2)}\right].$$

We can also find a confidence band for $\mu(x) = \alpha + \beta(x - \overline{x})$. The endpoints for a $100(1 - \alpha)\%$ confidence interval for $\mu(x)$ for a given $x$ are

$$\widehat{\alpha} + \widehat{\beta}(x - \overline{x}) \pm t_{\alpha/2}(n-2)\sqrt{\frac{n\widehat{\sigma^2}}{n-2}}\sqrt{\frac{1}{n} + \frac{(x - \overline{x})^2}{\sum_{i=1}^{n}(x_i - \overline{x})^2}}.$$

A prediction interval for a future value of $Y$, say $Y_{n+1}$, or a $100(1 - \alpha)\%$ confidence band for the observations of $Y$ is given by

$$\widehat{\alpha} + \widehat{\beta}(x - \overline{x}) \pm t_{\alpha/2}(n-2)\sqrt{\frac{n\widehat{\sigma^2}}{n-2}}\sqrt{1 + \frac{1}{n} + \frac{(x - \overline{x})^2}{\sum_{i=1}^{n}(x_i - \overline{x})^2}}.$$

## EXERCISES

**Purpose:** The exercises illustrate how the computer can be used to solve regression analysis problems. Graphical displays provide insight into the data.

**8.3–1** Using an Instron 4204, rectangular strips of plexi-glass were stretched to failure in a tensile test. The following data give the change in length in mm before breaking ($x$) and the cross-sectional area in mm$^2$($y$).

$$\begin{array}{ccccc} (5.28, 52.36) & (5.40, 52.58) & (4.65, 51.07) & (4.76, 52.28) & (5.55, 53.02) \\ (5.73, 52.10) & (5.84, 52.61) & (4.97, 52.21) & (5.50, 52.39) & (6.24, 53.77) \end{array}$$

(a) Plot the points and the least squares regression line on the same graph.

(b) Use `Residuals` to find and graph the residuals. Does it look like linear regression is appropriate? Why?

(c) Find and graph a 90% confidence band for $E(Y|x) = \mu(x)$.

(d) Find and graph a 90% prediction band for future observed values of $Y$.

**8.3–2** (The following information comes from the Westview Blueberry Farm and the National Oceanic and Atmospheric Administration Reports [NOAA].) For the paired data, $(x, y)$, $x$ gives the Holland rainfall for June and $y$ gives the blueberry production in thousands of pounds from the Westview Blueberry Farm. The data come from the years 1971 to 1989 for those years in which the last frost occurred May 10 or earlier.

$$\begin{array}{cccc} (4.11, 56.2) & (5.49, 45.3) & (5.35, 31.0) & (6.53, 30.1) \\ (5.18, 40.0) & (4.89, 38.5) & (2.09, 50.0) & (1.40, 45.8) \\ (4.52, 45.9) & (1.11, 32.4) & (0.60, 18.2) & (3.80, 56.1) \end{array}$$

(a) Plot the points and the least squares linear regression line on the same graph.

(b) Use `Residuals` to find and graph the residuals. Does it look like linear regression is appropriate? Why?

(c) Plot the points and the least squares quadratic regression line on the same graph.

(d) Find and graph the residuals for quadratic regression. Is this fit better than that shown by parts (a) and (b)?

**8.3–3** In a package of peanut $m$ & $m$'s, let $X$ equal the number of pieces of candy and $Y$ the total weight of the candies. The following data were observed:

| $x$ | $y$ | $x$ | $y$ |
|---|---|---|---|
| 19 | 46.5 | 19 | 47.2 |
| 20 | 49.8 | 20 | 48.1 |
| 20 | 50.3 | 20 | 49.1 |
| 20 | 48.8 | 20 | 51.6 |
| 20 | 50.5 | 21 | 50.2 |
| 21 | 50.0 | 21 | 52.2 |
| 21 | 49.7 | 21 | 51.6 |
| 21 | 52.5 | 21 | 53.5 |
| 22 | 52.4 | 22 | 52.9 |
| 22 | 51.1 | 22 | 54.4 |
| 22 | 53.4 | 22 | 52.7 |
| 22 | 51.3 | 23 | 52.9 |
| 23 | 54.8 | | |

(a) Plot the points and the least squares regression line on the same graph.

(b) Find and graph the residuals. Does it look like linear regression is appropriate? Why?

(c) Find and graph a 90% confidence band for $E(Y|x) = \mu_{Y|x}$.

(d) Find and graph a 90% prediction band for future observed values of $Y$.

(e) Let $X_{20}, X_{21}$, and $X_{22}$ equal the respective weights of bags of *m* & *m*'s containing 20, 21, and 22 pieces of candy. Assume that the respective distributions are $N(\mu_m, \sigma^2)$, $m = 20, 21, 22$. Use analysis of variance to test the null hypothesis $H_0: \mu_{20} = \mu_{21} = \mu_{22}$. You may select the significance level. Clearly state your conclusion and reason for it.

(f) Draw three box-and-whisker diagrams for the last part. Does this confirm your conclusion? Why?

(g) Test the null hypothesis $H_0\colon \mu_{20} = \mu_{22}$ against an appropriate alternative hypothesis. Calculate the $p$-value of your test. What is your conclusion and why?

**8.3–4** According to *Parade*, March 13, 1994, and the National Hospital Discharge Survey, 1991, the numbers of heart surgeries in thousands from 1980 to 1991 were as follows:

$$\begin{array}{cccc} (1980, 196) & (1981, 217) & (1982, 243) & (1983, 275) \\ (1984, 314) & (1985, 379) & (1986, 490) & (1987, 588) \\ (1988, 674) & (1989, 719) & (1990, 781) & (1991, 839) \end{array}$$

(a) Plot the points and the linear least squares regression line.

(b) Make a graph of the 12 residuals.

(c) From the residual plot, does linear regression seem to be appropriate?

(d) Either try cubic regression or a regression curve of your choice. Support your choice by comparing residual plots.

**8.3–5** The following data were reported in the August, 1991, issue of *HIV/AIDS Surveillance* published by the Centers for Disease Control. The Cases column gives the number of AIDS cases (not just HIV infections) diagnosed in the designated interval.

| Year | Months | Cases |
|---|---|---|
| 1981 | Jan-June | 92 |
| | July-Dec | 203 |
| 1982 | Jan-June | 390 |
| | July-Dec | 689 |
| 1983 | Jan-June | 1,277 |
| | July-Dec | 1,642 |
| 1984 | Jan-June | 2,550 |
| | July-Dec | 3,368 |
| 1985 | Jan-June | 4,842 |
| | July-Dec | 6,225 |
| 1986 | Jan-June | 8,215 |
| | July-Dec | 9,860 |
| 1987 | Jan-June | 12,764 |
| | July-Dec | 14,173 |
| 1988 | Jan-June | 16,113 |
| | July-Dec | 16,507 |
| 1989 | Jan-June | 18,452 |
| | July-Dec | 18,252 |
| 1990 | Jan-June | 18,601 |
| | July-Dec | 16,636 |
| 1991 | Jan-June | 12,620 |

(Note that the last two numbers reported are probably too low because of delayed reporting. You should be able to find updated figures.)

(a) Plot the first 19 points and the linear least squares regression line.

(b) Make a graph of the residuals.

(c) From the residual plot, does linear regression seem to be appropriate?

(d) Either try cubic regression or a regression curve of your choice. Support your choice by comparing residual plots.

# Chapter 9

# Multivariate Distributions

## 9.1 The Correlation Coefficient

Let $X$ and $Y$ have a bivariate distribution with p.d.f. $f(x,y)$. Then

- $\mu_X = E(X)$ is the mean of $X$,
- $\mu_Y = E(Y)$ is the mean of $Y$,
- $\sigma_X^2 = E[(X-\mu_X)^2]$ is the variance of $X$,
- $\sigma_Y^2 = E[(Y-\mu_Y)^2]$ is the variance of $Y$,
- $\sigma_{XY} = E[(X-\mu_X)(Y-\mu_Y)]$ is the covariance of $X$ and $Y$,
- $\rho_{XY} = \dfrac{\sigma_{XY}}{\sigma_X \sigma_Y}$ is the correlation coefficient of $X$ and $Y$.

The correlation coefficient, $\rho = \rho_{XY}$, gives a measure of the linear relationship between $X$ and $Y$. It is always true that $-1 \le \rho \le 1$.

The least squares regression line is given by

$$y = \mu_Y + \rho \frac{\sigma_Y}{\sigma_X}(x - \mu_X).$$

A random experiment, which is associated with a pair of random variables $(X, Y)$, is repeated $n$ times. Let $(X_1, Y_1), (X_2, Y_2), \ldots, (X_n, Y_n)$ denote the $n$ pairs of random variables associated with the $n$ trials. These $n$ pairs of random variables are called a random sample of size $n$ from the distribution associated with $(X, Y)$. For the observed values of this random sample, $(x_1, y_1), (x_2, y_2), \ldots, (x_n, y_n)$,

- $\overline{x} = \dfrac{1}{n}\sum_{i=1}^{n} x_i$ is the sample mean of the $x_i$'s,

- $\overline{y} = \frac{1}{n}\sum_{i=1}^{n} y_i$ is the sample mean of the $y_i$'s,

- $s_x^2 = \frac{1}{n-1}\sum_{i=1}^{n}(x_i - \overline{x})^2$ is the sample variance of the $x_i$'s,

- $s_y^2 = \frac{1}{n-1}\sum_{i=1}^{n}(y_i - \overline{y})^2$ is the sample variance of the $y_i$'s,

- $r = \dfrac{\frac{1}{n-1}\sum_{i=1}^{n}(x_i - \overline{x})(y_i - \overline{y})}{s_x s_y}$ is the sample correlation coefficient.

The observed least squares regression line or "best fitting line" is

$$\widehat{y} = \overline{y} + r\left(\frac{s_y}{s_x}\right)(x - \overline{x})$$

where $\overline{x}$, $\overline{y}$, $s_x$, $s_y$, and $r$ are the observed values calculated from the data.

The difference, $y_i - \widehat{y}_i$, is called the $i$th residual, $i = 1, 2, \ldots, n$. The sum of the residuals should always equal zero. A residual scatterplot of the points $(x_i, y_i - \widehat{y}_i)$, often provides insight as to whether or not a linear regression line is the most appropriate "best fitting line." (See Exercise 9.2-12.)

## EXERCISES

**Purpose:** The exercises illustrate the relation between distribution and sample characteristics.

**9.1–1** A pair of 8-sided dice is rolled. Let $X$ denote the smaller and $Y$ the larger outcome on the dice.

(a) Simulate 50 repetitions of this experiment.

(b) Use `ScatPlotLine(X,Y)` to plot these points along with the least squares regression line.

(c) Compare the sample characteristics with the distribution characteristics (e.g., means, variances). Are they approximately equal? How would a larger sample size affect the closeness of the sample and distribution characteristics?

**9.1–2** Roll 9 fair 4-sided dice. Let $X$ equal the number of outcomes that equal 1. Let $Y$ equal the number of outcomes that equal 2 or 3.

(a) Simulate 50 repetitions of this experiment.

(b) Use `ScatPlotLine(X,Y)` to plot these points along with the least squares regression line.

(c) Compare the sample characteristics with the distribution characteristics. Are they approximately equal? How would a larger sample affect the closeness of the sample and distribution characteristics?

**9.1–3** Roll 9 fair 4-sided dice. Let $X$ equal the number of outcomes that equal 1. Let $Y$ equal the number of outcomes that equal 2 or 3.

(a) Define the joint p.d.f. of $X$ and $Y$ and draw a probability histogram of this p.d.f. For example, do the following:

```
f := (x,y) -> binomial(9,x)*binomial(9-x,y)*(1/4)^x*
(1/2)^y*(1/4)^(9-x-y);
P := [seq([seq(f(i,j),j = 0 .. 9)],i = 0 .. 9)];
with(plots):
matrixplot(P,heights = histogram);
```

(b) Simulate 1000 repetitions of this experiment. Construct a relative frequency histogram of your joint data. Here is one way to do this:

```
m := 1000:
Freqs := [seq(Freq(Die(4,9),1 .. 4),i = 1 .. m)]:
X := [seq(F[1],F = Freqs)]:
Y := [seq(F[2]+F[3],F = Freqs)]:
for i from 0 to 9 do
for j from 0 to 9 do  freq[i][j] := 0 od
od:
for i from 1 to m do
freq[X[i]][Y[i]] := freq[X[i]][Y[i]]+1
od:
relfreq := [seq([seq(evalf(freq[i][j]/m),j = 0 .. 9)],
i = 0 .. 9)]:
with(plots):
matrixplot(relfreq,heights = histogram);
```

(c) Draw the relative frequency histogram of the 1000 observations of $X$ with the probability histogram for the $b(9, 0.25)$ distribution superimposed. How is the fit?

(d) Draw the relative frequency histogram of the 1000 observations of $Y$ with the probability histogram for the $b(9, 0.50)$ distribution superimposed. How is the fit?

(e) Visually compare the histograms for parts (a) and (b).

## 9.2 Conditional Distributions

Let $f(x, y)$ denote the joint p.d.f. of the random variables $X$ and $Y$. Let $f_1(x)$, $x \in R_1$, and $f_2(y)$, $y \in R_2$, denote the marginal p.d.f.'s of $X$ and $Y$, respectively.

The conditional p.d.f. of $X$, given $Y = y$, is defined by

$$g(x|y) = \frac{f(x,y)}{f_2(y)}, \quad x \in R_1,$$

provided that $f_2(y) > 0$.

The conditional p.d.f. of $Y$, given $X = x$, is defined by

$$h(y|x) = \frac{f(x,y)}{f_1(y)}, \quad y \in R_2,$$

provided that $f_1(x) > 0$.

Note that

$$f(x,y) = f_1(x)\, h(y|x) = f_2(y)\, g(x|y).$$

The random variables $X$ and $Y$ are independent in case

$$f(x,y) = f_1(x)\, f_2(y)$$

or

$$g(x|y) = f_1(x)$$

or

$$h(y|x) = f_2(y).$$

When $E(Y|x)$ is linear,

$$E(Y|x) = \mu_Y + \rho\frac{\sigma_Y}{\sigma_X}(x - \mu_X).$$

When $E(X|y)$ is linear,

$$E(X|y) = \mu_X + \rho\frac{\sigma_X}{\sigma_Y}(y - \mu_Y).$$

### EXERCISES

**Purpose:** The exercises use conditional distributions to simulate samples from joint distributions. In addition the relation between the least squares regression line and $E(Y|x)$ is illustrated.

**9.2–1** Let $X$ and $Y$ have a uniform distribution on the points with integer coordinates in

$$D = \{(x, y) : 1 \le x \le 8,\ x \le y \le x + 2\}.$$

That is $f(x, y) = 1/24$, $(x, y) \in D$, when $x$ and $y$ are both integers.

(a) Define the marginal p.d.f. of $X$, $f_1(x)$.

(b) Define the conditional p.d.f. of $Y$, given $X = x$, namely $h(y|x)$.

(c) Find the marginal p.d.f. of $Y$.

(d) Simulate a random sample of size 20 from this joint distribution.

(e) Define $E(Y|x)$ as a linear function.

(f) Use `ScatPlotLine(X,Y)` to plot these 20 observations. Superimpose the graph of $y = E(Y|x)$ from part (e). (Explain each of the following steps.)

```
X := Die(8,20):
B := Die(3,20):
Y := [seq(X[k]+B[k]-1,k = 1 .. 20)]:
plot({ScatPlotLine(X,Y),plot(x+1,x = 1 .. 8)});
```

(g) Are $X$ and $Y$ independent random variables? Verify your answer.

**9.2–2** Continuing with the last exercise, simulate 200 observations of $X$ and then, conditionally, simulate 200 values of $Y$. Construct a relative frequency histogram of the 200 observations of $Y$. If possible, superimpose the probability histogram of $Y$.

**9.2–3** In Exercise 9.1-3 we simulated observations of the random variables $X$ and $Y$ that had a trinomial distribution with $p_1 = 0.25$, $p_2 = 0.50$ and $n = 9$. In this exercise we shall do the simulation using conditional distributions.

(a) First simulate 200 observations of $X$, a binomial random variable that is $b(9, 0.25)$.

(b) Conditionally, given $X = x$, the distribution of $Y$ is $b(9 - x, p_2/(1 - p_1))$. So for each value of $X$, simulate an observation of $Y$ using its conditional distribution. Explain why the following will do this.

```
m := 200;
X := BinomialS(9,1/4,m);
Y := [seq(op(BinomialS(9-X[i],2/3,1)),i = 1 .. m)];
```

(c) Use `ScatPlotLine(X,Y)` to make a scatter plot of the observations of $X$ and $Y$, and superimpose the graph of $y = E(Y|x) = \mu_Y + \rho(\sigma_Y/\sigma_X)(x - \mu_X)$. How is the fit?

(d) Although the observations of $Y$ were simulated conditionally, the marginal distribution of $Y$ is $b(9, 0.50)$. Illustrate this by comparing sample characteristics with distribution characteristics. On the same figure, draw the graphs of the relative frequency histogram of the observations of $Y$ along with the probability histogram of $Y$. Also draw the graphs of the empirical and theoretical distribution functions for $Y$.

(e) As you did in Exercise 9.1-3(b), construct a relative frequency histogram of the joint data.

**9.2–4** Repeat the last exercise for other values of the parameters.

**9.2–5** Let the joint p.d.f. of the random variables $X$ and $Y$ be defined by $f(x, y) = 1/15,\ 1 \le x \le y \le 5$, where $x$ and $y$ are integers.

(a) Show that the marginal p.d.f. of $X$ is $f_1(x) = x/15,\ x = 1, 2, 3, 4, 5$.

(b) Show that the conditional p.d.f. of $Y$, given $X = x$, is $h(y|x) = 1/x,\ y = 1, 2, \ldots, x$. That is, the conditional distribution of $Y$, given $X = x$, is uniform on the integers $1, 2, \ldots, x$.

(c) Define $E(Y|x)$ as a linear function.

(d) Simulate a random sample of size 100 from this distribution by first generating an observation of $X$, and then, for $X = x$, generate an observation of $Y$.

(e) Use `ScatPlotLine` to obtain a scatter plot of these data along with the least squares regression line. Superimpose on your graph, $y = E(Y|x)$.

(f) Compare the equations of $E(Y|x)$ and the observed least squares regression line. Are they equal?

(g) Graph the relative frequency histogram of the observations of $Y$ with the probability histogram of the distribution of $Y$ superimposed. Also compare the sample mean and sample variance of the observations of $Y$ with $\mu_Y$ and $\sigma_Y^2$, respectively.

**9.2–6** Let $X$ have a uniform distribution on the integers $0, 1, 2, \ldots, 9$. Given that $X = x$, let $Y$ have a uniform distribution on the integers $x, x + 1, \ldots, 9$.

(a) Simulate a random sample of size 200 from the joint distribution.

(b) Use `MarginalRelFreq` to compare your empirical results with the theoretical distribution.

(c) Use `ScatPlotLine` to plot the observations that were simulated. Superimpose the conditional mean line, $y = E(Y|x)$? Is the conditional mean line close to the least squares regression line?

(d) Graph the relative frequency histogram of the observations of $Y$ with the probability histogram of the distribution of $Y$ superimposed. Also compare the sample mean and sample variance of the observations of $Y$ with $\mu_Y$ and $\sigma_Y^2$, respectively.

**9.2–7** Let the marginal p.d.f. of $X$ be $f_1(x) = 1,\ 0 < x < 1$. Let the conditional p.d.f. of $Y$, given $X = x$, be $h(y|x) = 1/(1-x),\ x < y < 1$, for $0 < x < 1$. That is, $Y$ is conditionally $U(x, 1)$.

(a) Define $f(x, y)$, the joint p.d.f. of $X$ and $Y$.

(b) Define $f_2(y)$, the marginal p.d.f. of $Y$.

(c) Simulate a random sample of size 200 from this bivariate distribution. First generate an observation $x$ of $X$ then, given $X = x$, generate an observation of $Y$.

(d) Obtain the correlation coefficient and regression line for data.

(e) Define $E(Y|x)$ as a linear function.

(f) Compare the equations of $E(Y|x)$ and the observed least squares regression line. Are they about equal?

(g) Use `ScatPlotLine` to plot a scatter diagram of the observations along with the observed least squares regression line. Superimpose the graph of $y = E(Y|x)$.

(h) Graph the relative frequency histogram of the observations of $Y$ with the probability histogram of the distribution of $Y$ superimposed. Also compare the sample mean and sample variance of the observations of $Y$ with $\mu_Y$ and $\sigma_Y^2$, respectively.

**9.2–8** Let $f(x, y) = 1/20,\ x < y < x + 2,\ 0 < x < 10$, be the joint p.d.f. of $X$ and $Y$.

(a) Find $f_1(x)$, the marginal p.d.f. of $X$.

(b) Find $h(y|x)$, the conditional p.d.f. of $Y$, given $X = x$.

(c) What does $E(Y|x)$ equal?

(d) Generate a random sample of size 100 from this bivariate distribution.

(e) Use `ScatPlotLine` to plot a scatter diagram of the observations along with the observed least squares regression line. Superimpose the graph of $y = E(Y|x)$.

(f) Compare the equations of $E(Y|x)$ and the observed least squares regression line. Are they about equal?

(g) Graph the relative frequency histogram of the observations of $Y$ with the probability histogram of the distribution of $Y$ superimposed. Also compare the sample mean and sample variance of the observations of $Y$ with $\mu_Y$ and $\sigma_Y^2$, respectively.

**9.2–9** Let $f(x, y) = 2$ for $0 \le x \le y \le 1$ be the joint p.d.f. of $X$ and $Y$.

(a) Find the marginal p.d.f.'s of $X$ and $Y$.

(b) Simulate 200 observations of the pair of random variables $(X, Y)$. Do this by simulating pairs of random numbers and let $X$ equal the smaller and $Y$ the larger of the pair.

(c) Plot the pairs of points using `ScatPlotLine`, superimposing $y = E(Y|x) = (x + 1)/2$.

(d) Plot a histogram of the observations of $X$ with the marginal p.d.f. of $X$ superimposed.

(e) Plot a histogram of the observations of $Y$ with the marginal p.d.f. of $Y$ superimposed.

**9.2–10** Let a number $X$ be selected at random from the interval $(0, 10)$. For each observed value of $X = x$, let a number $Y$ be selected at random from the interval $(-x^2 + 10x,\ -x^2 + 10x + 5)$. Simulate 50 observations of the pair of random variables $(X, Y)$.

(a) First plot your set of data using `ScatPlot`.

(b) Now plot your data with a quadratic regression line. **Hint:** Given that the data are stored in lists `X` and `Y`, the quadratic regression line is given by `yhat := PolyReg(X,Y,2,x);`.

(c) Superimpose over your last graph the graph of $y = E(Y|x)$ and interpret your output.

(d) What is the value of the sample correlation coefficient?

**9.2–11** Change the last exercise by selecting $Y$ randomly, for a given $X = x$, from the interval $(x^2 - 10x + 25,\ x^2 - 10x + 30)$.

**9.2–12** A pharmaceutical company is interested in the stability of vitamin $B_6$ in a multiple vitamin pill that it manufactures. They plan to manufacture the pills with 50 milligrams of vitamin $B_6$ per pill. Twenty pills were selected at random at each of five different times after the manufactured date and the amount of vitamin $B_6$ was determined. Table 9.2–1 lists these observations along with the the pill was manufactured.

(a) For a specific time period $X = x$ $(x = 0, 1, 2, 3, 4)$, let $Y$ be a randomly chosen observation of vitamin $B_6$ content for that $x$. Obtain a scatter plot with the least squares linear regression line for these data. (The data are stored in the file `STAT.DAT` under the names `D9212X` and `D2121Y`.)

(b) Make a scatter plot of the residuals. Does linear regression seem to be appropriate?

(c) Obtain a scatter plot with the least squares quadratic curve included. Again make a scatter plot of the residuals.

(d) Does linear or quadratic regression seem to be more appropriate? Base your answer on the output from parts (b) and (c).

| Time | | | | | |
|---|---|---|---|---|---|
| 0 | 1 | 2 | 3 | 4 | |
| 1 | 49.75 | 49.15 | 48.15 | 46.60 | 45.60 |
| 2 | 49.60 | 49.35 | 48.30 | 46.95 | 45.70 |
| 3 | 49.35 | 49.30 | 48.20 | 47.10 | 45.50 |
| 4 | 49.15 | 49.35 | 48.25 | 47.00 | 45.35 |
| 5 | 49.45 | 49.25 | 48.30 | 46.90 | 45.45 |
| 6 | 49.50 | 49.30 | 48.25 | 47.00 | 45.60 |
| 7 | 49.60 | 49.15 | 48.35 | 46.60 | 45.70 |
| 8 | 49.70 | 49.05 | 48.05 | 46.95 | 45.50 |
| 9 | 49.55 | 49.20 | 48.45 | 46.85 | 45.45 |
| 10 | 49.30 | 49.25 | 48.30 | 46.90 | 45.25 |
| 11 | 49.35 | 49.10 | 48.20 | 47.05 | 45.35 |
| 12 | 49.25 | 49.00 | 48.15 | 46.95 | 45.45 |
| 13 | 49.85 | 49.15 | 48.25 | 46.85 | 45.80 |
| 14 | 49.50 | 49.20 | 48.25 | 47.05 | 45.60 |
| 15 | 49.30 | 49.25 | 48.35 | 47.15 | 45.55 |
| 16 | 49.35 | 49.50 | 48.10 | 46.95 | 45.50 |
| 17 | 49.65 | 48.95 | 48.30 | 46.90 | 45.70 |
| 18 | 49.55 | 49.25 | 48.25 | 46.90 | 45.60 |
| 19 | 49.50 | 49.30 | 48.20 | 47.00 | 45.50 |
| 20 | 49.65 | 49.30 | 48.30 | 47.15 | 45.60 |

**Table 9.2–1**

## Questions and Comments

**9.2–1** If you omitted Exercises 2.1-4 and 2.1-5, this would be an appropriate time to work them.

# 9.3 The Bivariate Normal Distribution

Let $X$ and $Y$ have a bivariate normal distribution with parameters $\mu_X$, $\sigma_X^2$, $\mu_Y$, $\sigma_Y^2$, and $\rho$. Then

1. The marginal distribution of $X$ is $N(\mu_X, \sigma_X^2)$.

2. The conditional distribution of $Y$, given $X = x$, is

$$N(\mu_Y + \rho \frac{\sigma_Y}{\sigma_X} [x - \mu_X],\ \sigma_Y^2 [1 - \rho^2]).$$

3. $E(Y|x) = \mu_X + \rho\,[\sigma_Y/\sigma_X][x - \mu_X]$.

4. The marginal distribution of $Y$ is $N(\mu_Y, \sigma_Y^2)$.

## EXERCISES

**Purpose:** The exercises illustrate empirically the above four properties.

**9.3–1** Let $X$ and $Y$ have a bivariate normal distribution with parameters $\mu_X = 50$, $\sigma_X^2 = 36$, $\mu_Y = 70$, $\sigma_Y^2 = 64$, and $\rho = 0.80$.

(a) Simulate a random sample of size 100 from this distribution as follows: First simulate an observation of $X$ from the normal distribution $N(50, 36)$ and then for an observed value, $x$, simulate an observation of $Y$ using the conditional distribution of $Y$, given $X = x$.

(b) Find the correlation coefficient and least squares regression line of the data and compare these to their theoretic counterparts.

(c) Plot the paired data with its least squares regression line and superimpose the graph of $y = E(Y|x)$. How is the fit?

(d) Make a histogram of the observations of $Y$ and superimpose the p.d.f. for the $N(70, 64)$ distribution.

**9.3–2** Repeat Exercise 9.3-1 for other values of $\rho$. For example, let $\rho = -0.80$, $-0.25$, and/or $0.25$.

**9.3–3** Make a graph of the joint p.d.f. of $X$ and $Y$ for each of the distributions that you sampled from in the last two exercises. Also make a graph of the contours and note their elliptical shape. Make the appropriate changes in the following statements.

```
mux := 50;
muy := 70;
varx := 36;
vary := 64;
rho := .80;
f := BivariateNormalPDF(mux,varx,x,muy,vary,y,rho);
plot3d(f,x = 35 .. 65,y = 50 .. 100,numpoints = 3000);
```

## Questions and Comments

**9.3–1** In the future, use `BivariateNormalS` to simulate bivariate normal data.

# 9.4 Correlation Analysis

Let $(X_1, Y_1), (X_2, Y_2), \ldots, (X_n, Y_n)$ be a random sample of size $n$ from a bivariate normal distribution. Let

$$R = \frac{\frac{1}{n-1}\sum_{i=1}^{n}(X_i - \overline{X})(Y_i - \overline{Y})}{\sqrt{\frac{1}{n-1}\sum_{i=1}^{n}(X_i - \overline{X})^2}\sqrt{\frac{1}{n-1}\sum_{i=1}^{n}(Y_i - \overline{Y})^2}} = \frac{S_{XY}}{S_X S_Y}.$$

When $\rho = 0$,

$$T = \frac{R\sqrt{n-2}}{\sqrt{1-R^2}}$$

has a Student's $t$ distribution with $r = n - 2$ degrees of freedom.

When $\rho = 0$, the distribution function and p.d.f. of $R$ are given, respectively, by

$$G(r) = P(R \le r) = P(T \le r\sqrt{n-2}/\sqrt{1-r^2})$$

and

$$g(r) = \frac{\Gamma[(n-1)/2]}{\Gamma(1/2)\,\Gamma[(n-2)/2]}(1-r^2)^{(n-4)/2}, \quad -1 \le r \le 1.$$

When $n$ is sufficiently large,

$$W_n = \frac{(1/2)\ln\left(\dfrac{1+R}{1-R}\right) - (1/2)\ln\left(\dfrac{1+\rho}{1-\rho}\right)}{\sqrt{1/(n-3)}}$$

has a distribution which is approximately $N(0,1)$.

## EXERCISES

**Purpose:** The exercises investigate the distribution of $R$ and the approximate distribution of $W_n$.

**9.4–1** Graph and interpret the p.d.f.'s of $R$ when $\rho = 0$ and $n = 3, 4, 5, 6,$ and 7. Do these graphs surprise you?

**9.4–2** Graph and interpret the p.d.f.'s of $R$ when $\rho = 0$ and $n = 10, 15,$ and 20.

**9.4–3** Find the mean and the variance of the distribution of $R$ when $\rho = 0$..

**9.4–4** Let $(X_1, Y_1), (X_2, Y_2), \ldots, (X_n, Y_n)$ be a random sample of size $n$ from independent normal distributions, $N(\mu_X, \sigma_X^2)$ and $N(\mu_Y, \sigma_Y^2)$. Let $R$ denote the sample correlation coefficient.

(a) For $n = 3, 4$, and 5, simulate 100 (or 200) observations of $R$. You may choose values for the means and variances of the normal distributions. Note that `Correlation(X,Y);` provides the correlation coefficient of the two lists of observations in `X` and in `Y`.

(b) Construct a relative frequency histogram of your data with the appropriate p.d.f. superimposed.

(c) Plot the empirical distribution function of your data with the distribution function of $R$ superimposed.

**9.4–5** Let $(X_1, Y_1), (X_2, Y_2), \ldots, (X_n, Y_n)$ be a random sample of size $n$ from independent normal distributions, $N(\mu_X, \sigma_X^2)$ and $N(\mu_Y, \sigma_Y^2)$. Let $R$ denote the sample correlation coefficient.

(a) For $n = 3, 4$, and 5, simulate 100 (or 200) observations of $R$. You may choose values for the means and variances of the normal distributions.

(b) For each observation of $R$, calculate the value of $T = R\sqrt{n-2}/\sqrt{1-R^2}$.

(c) Plot a relative frequency histogram of the observations of $T$ with the p.d.f. of $T$ superimposed. Be sure to specify the proper number of degrees of freedom.

(d) Plot the ogive curve of the observations of $T$ with the distribution function of $T$ superimposed.

**9.4–6** Test the hypothesis $H_0$: $W_n$ is $N(0, 1)$ when $n = 5$. In particular, generate 100 or 200 observations of $W_5$. (Note that $W_n$ is defined earlier in this section.)

(a) Plot a relative frequency histogram of the observations of $W_5$ with the $N(0, 1)$ p.d.f. superimposed.

(b) Use the Kolmogorov-Smirnov goodness of fit test. (See Section 10.7.)

**9.4–7** Repeat the last exercise for $n = 10$ and for another value of $n$.

## Questions and Comments

**9.4–1** For what values of $n$ is $n$ sufficiently large so that the distribution of $W_n$ is approximately $N(0, 1)$?

## 9.5 The Basic Chi-Square Statistic

Let $W_1, W_2, \ldots, W_n, \ldots$ be a sequence of random variables; let $F(w;n)$ be the distribution function of $W_n$; let $F(w)$ be the distribution function of $W$. The sequence $\{W_n\}$ converges to $W$ in distribution if

$$\lim_{n\to\infty} F(w;n) = F(w)$$

at every point of continuity of $F(w)$.

We shall illustrate convergence in distribution with the following example illustrating the basic chi-square statistic.

Consider a sequence of repetitions of an experiment for which the following conditions are satisfied:

1. The experiment has $k$ possible outcomes that are mutually exclusive and exhaustive, say, $A_1, A_2, \ldots, A_k$;

2. $n$ independent trials of this experiment are observed;

3. $P(A_i) = p_i$, $i = 1, 2, \ldots, k$, on each trial with $\sum_{i=1}^{k} p_i = 1$.

Let the random variable $Y_i$ be the number of times $A_i$ occurs in the $n$ trials, $i = 1, 2, \ldots, k$. If $y_1, y_2, \ldots, y_k$ are nonnegative integers such that their sum equals $n$, then for such a sequence, the probability that $A_i$ occurs $y_i$ times, $i$ = 1,2,. . .,$k$, is given by

$$P(Y_1 = y_1, \cdots, Y_k = y_k) = \frac{n!}{y_1! \cdots y_k!} p_1^{y_1} \cdots p_k^{y_k}.$$

The random variables $Y_1, Y_2, \ldots, Y_k$ have a multinomial distribution and $E(Y_i) = np_i$, $i = 1, 2, \ldots, k$.

Let

$$Q_{k-1} = \sum_{i=1}^{k} \frac{(Y_i - np_i)^2}{np_i}$$

where $Y_i$ denotes the observed number of occurrences of $A_i$ and $np_i$ is the expected number of occurrences of $A_i$. We shall illustrate empirically that $Q_{k-1}$ converges in distribution to a random variable $Q$, where $Q$ is $\chi^2(k-1)$. [Perhaps we should let $Q_{k-1} = (Q_{k-1})_n$ to emphasize that the distribution of $Q_{k-1}$ depends on $n$. However, we shall not do this.]

Because $Q_{k-1}$ converges in distribution to $Q$, it should be true that, for $n$ sufficiently large, $Q_{k-1}$ has a distribution that is approximately $\chi^2(k-1)$. The "rule of thumb" that is often given is that $np_i$ should be greater than 5 for all $i$. This "rule of thumb" will be tested.

The statistic $Q_{k-1}$ is the basic chi-square statistic. A variety of tests, including goodness of fit tests, are based on this statistic.

## EXERCISES

**Purpose:** The exercises illustrate empirically that $Q_{k-1}$ is approximately $\chi^2(k-1)$ when $n$ is "sufficiently large."

**9.5–1** Let $A_1 = [0, 0.2)$, $A_2 = [0.2, 0.4)$, $A_3 = [0.4, 0.6)$, $A_4 = [0.6, 0.8)$, and $A_5 = [0.8, 1.0)$.

(a) Generate $n = 10$ random numbers. Let $y_i$ equal the number of times that $A_i$ occurs, $i = 1, 2, 3, 4, 5$. Since $p_i = 1/5$, $np_i = 2$ and it follows that

$$q_{k-1} = q_4 = \sum_{i=1}^{5} \frac{(y_i - 2)^2}{2}.$$

Calculate 50 observations of $q_4$, each based on a random sample of 10 random numbers. Plot the empirical distribution function of these 50 observations of $q_4$ along with the theoretical distribution function for the chi-square distribution $\chi^2(4)$.

(b) Use the Kolmogorov-Smirnov goodness of fit statistic to test the hypothesis $H_0$: $Q_4$ is $\chi^2(4)$. Let $\alpha = 0.10$. Find the sample mean and sample variance of the $q_4$'s.

(c) Repeat parts (a) and (b) for $n = 5, 15$, and 25. For what values of $np$ would you accept the hypothesis that $Q_4$ has a chi-square distribution with 4 degrees of freedom?

**9.5–2** Let $A_i = [(i-1)/10, i/10)$, $i = 1, 2, \ldots, 10$.

(a) Generate $n = 20$ random numbers. Let $y_i$ equal the number of times that $A_i$ occurs, $i = 1, 2, \ldots, 10$. Since $p_i = 1/10$, $np_i = 2$ and

$$q_{k-1} = q_9 = \sum_{i=1}^{10} \frac{(y_i - 2)^2}{2}.$$

Calculate 50 observations of $q_9$, each based on a random sample of 20 random numbers. Plot the empirical distribution function of these 50 observations of $q_9$ along with the theoretical distribution function for the chi–square distribution $\chi^2(9)$.

(b) Test the hypothesis $H_0$: $Q_9$ is $\chi^2(9)$ using the Kolmogorov-Smirnov statistic. Let $\alpha = 0.10$.

(c) Repeat parts (a) and (b) for $n = 10, 30$, and 50. For what values of $np$ would you accept the hypothesis that $Q_9$ has a chi–square distribution with 9 degrees of freedom?

## Questions and Comments

**9.5–1** In Exercise 9.5–1, plot a relative frequency histogram of 100 observations of $Q_4$ when $n = 25$. Superimpose the $\chi^2(4)$ p.d.f.

**9.5–2** Repeat Exercise 9.5–2 with sets of unequal size. For example, let $A_1 = [0, 0.05)$, $A_2 = [0.05, 0.20)$, etc. Or let $A_1 = [0.0, .01)$, $A_2 = [0.01, 0.20)$, etc. Are the results of this experiment much different from those of Exercise 9.5–2?

**9.5–3** On the basis of your observations, how large must $np_i$ be in order for $Q_{k-1}$ to be approximately $\chi^2(k-1)$.

# 9.6 Testing Probabilistic Models

Let an experiment have $k$ mutually exclusive and exhaustive outcomes, $A_1, A_2, \ldots, A_k$. Let $p_i = P(A_i)$. To test the hypothesis $H_0: p_i = p_{i0}$, $p_{i0}$ known, $i = 1, 2, \ldots, k$, repeat the experiment $n$ independent times. Let $Y_i$ equal the observed number of times that $A_i$ occurs. Then

$$Q_{k-1} = \sum_{i=1}^{k} \frac{(Y_i - np_{i0})^2}{np_{i0}}$$

is approximately $\chi^2(k–1)$ when $H_0$ is true. The hypothesis $H_0$ is rejected if $q_{k-1} \geq \chi^2_\alpha(k–1)$ for an $\alpha$ significance level.

If the probabilities $p_{10}, p_{20}, \ldots, p_{k0}$ are functions of one or more unknown parameters, one degree of freedom is lost for each parameter that is estimated.

Let $W$ be a random variable of the continuous type with distribution function $F(w)$. We wish to test $H_0: F(w) = F_0(w)$ where $F_0(w)$ is some unknown distribution function. In order to use the chi–square statistic, partition the set of possible values of $W$ into $k$ sets, say $A_1, A_2, \ldots, A_k$. Let $p_i = P(W \in A_i)$ and let $p_{i0} = P(W \in A_i)$ when the distribution function of W is $F_0(w)$. The hypothesis that is actually tested is $H_0': p_i = p_{i0}$, $i = 1, 2, \ldots, k$. If $H_0'$ is not rejected, then $H_0$ is not rejected.

## EXERCISES

**Purpose:** The chi-square goodness of fit test is used to test some of the procedures that are provided to simulate samples from various distributions. Determining whether data fit a probability model is examined. The loss of degrees of freedom when estimating unknown parameters is illustrated.

**9.6–1** To help understand the chi-square goodness of fit test procedure, simulate 200 observations of a discrete uniform distribution on $1, 2, \ldots, 10$. Then the theoretical distribution function is $F := x/10$ for $x = 1, 2, \ldots, 10$. The procedure `ChisquareFit` provides both a graphical comparison as well as the value of the chi-square goodness of fit statistic.

```
X := Die(10,200);
ChisquareFit(X,x/10,[$(1 .. 11)],Discrete);
```

**9.6–2** (a) Use `X := BinomialS(4,1/2,200);` to generate 200 observations of a random variable $X$ where $X$ is $b(4, 1/2)$.

(b) In order to test the hypothesis that $X$ is $b(4, 1/2)$, use these data with `F := BinomialCDF(4,0.5,x);` and `ChisquareFit(X,F,[$0 .. 5],Discrete);`. That is, test whether `BinomialS` performs as expected.

**9.6–3** (a) Use `X := PoissonS(4,200);` to generate 200 observations of a Poisson random variable, $X$, with a mean of $\lambda = 4$.

(b) To test the hypothesis that $X$ is Poisson, $\lambda = 4$, use these data, `F := PoissonCDF(4,x);` and `ChisquareFit(X,F,[$(0..Max(X)+1)],Discrete);` Again we are really testing how well `PoissonS` performs.

**9.6–4** It is claimed by Hoel (*Introduction to Mathematical Statistics*, fifth edition, pp. 94-95) that "the number of deaths from the kick of a horse per army corps per year, for 10 Prussion Army Corps for 20 years" has a Poisson distribution. The data, given as a list of lists [number of deaths, frequency] are

```
L := [[0, 109], [1, 65], [2, 22], [3, 3], [4, 1]];
```

Use `ChisquareFit` to test whether these data are observations of a Poisson random variable. Note that one degree of freedom is lost because $\lambda$ must be estimated from the data.

**9.6–5** Use `ChisquareFit` to test whether data sets of discrete data in your text book are observations of a distribution claimed for them. Or, alternatively, test whether the data in Table 3.7-1, Exercise 3.7-6, gives observations of a Poisson random variable. (This is data set `D376` in `STAT.DAT`.)

**9.6–6** (a) Use `ExponentialS` with $\theta = 10$ to generate 200 observations of an exponentially distributed random variable $X$ with mean $\theta = 10$.

(b) Use these data and the chi-square statistic to test the hypothesis that $X$ has an exponential distribution with mean $\theta = 10$. That is, test whether `ExponentialS` performs as expected.

## Questions and Comments

**9.6–1** In Section 5.4, if you did not use the chi–square goodness of fit test when working the Central Limit Theorem problems, do so now.

**9.6–2** You could modify Exercise 9.6–6 to illustrate the loss of 2 degrees of freedom when two parameters are estimated. For example, sample from a normal distribution. Calculate the chi–square values using the known mean, $\mu$, and variance $\sigma^2$, and then using estimates of $\mu$ and $\sigma^2$.

**9.6–3** In Exercise 9.6–6, when the parameter $\theta$ was estimated, the resulting chi–square statistic was, on the average, smaller than when $\theta = 1$ was used. Does this make sense to you intuitively? Explain.

# 9.7 Tests of the Equality of Multinomial Distributions

Consider two multinomial distributions with parameters $n_j$, $p_{1j}$, $p_{2j}, \ldots, p_{kj}$, $j = 1, 2$, respectively. Let $X_{ij}$, $i = 1, 2, \ldots, k$, $j = 1, 2$, represent the corresponding frequencies. We are interested in testing the null hypothesis

$$H_0\colon p_{11} = p_{12},\ p_{21} = p_{22}, \cdots,\ p_{k1} = p_{k2}$$

where each $p_{i1} = p_{i2}$, $i$ = 1,2,…,$k$, is unspecified. The statistic

$$Q = \sum_{j=1}^{2}\sum_{i=1}^{k} \frac{\{X_{ij} - n_j[(X_{i1} + X_{i2})/(n_1 + n_2)]\}^2}{n_j[(X_{i1} + X_{i2})/(n_1 + n_2)]}$$

has an approximate chi-square distribution with $r = 2k - 2 - (k - 1) = k - 1$ degrees of freedom. The null hypothesis is rejected if $q \geq \chi^2_\alpha(k-1)$.

Consider a random experiment, the outcome of which can be classified by two different attributes, the first having $k$ and the second $h$ mutually exclusive and exhaustive events, $A_1, \ldots, A_k$ and $B_1, \ldots, B_h$, respectively. We wish to test the hypothesis that the attributes of classification are independent, namely, $H_0:\ P(A_i \cap B_j) = P(A_i)P(B_j)$, $i = 1, 2, \ldots, k$, $j = 1, 2, \ldots, h$.

Given $n$ observations, let $Y_{ij}$, $Y_{i\cdot}$ and $Y_{\cdot j}$ denote the observed frequencies in $A_i \cap B_j$, $A_i$, and $B_j$, respectively. The random variable

$$Q = \sum_{j=1}^{h}\sum_{i=1}^{k} \frac{[Y_{ij} - n(Y_{i\cdot}/n)(Y_{\cdot j}/n)]^2}{n(Y_{i\cdot}/n)(Y_{\cdot j}/n)}$$

has an approximate chi-square distribution with $(k - 1)(h - 1)$ degrees of freedom when $H_0$ is true. The hypothesis $H_0$ is rejected if $q \geq \chi^2_\alpha([k-1][h-1])$.

## EXERCISES

**Purpose:** Some exercises illustrate empirically the distribution of $Q$ and the loss of degrees of freedom. Others illustrate the use of the computer for contingency table problems.

**9.7–1** Suppose that two bowls each contain chips numbered $1, 2, 3, 4$, or 5, one of each number. Let $p_{ij}$ denote the probability of drawing a chip numbered $i$ from bowl $j, i = 1, 2, 3, 4, 5,\ j = 1, 2$. To test the hypothesis $H_0 : p_{11} = p_{12}, \cdots, p_{51} = p_{52}$, the statistic $Q$ defined above can be used. Note that $Q$ has 4 degrees of freedom.

(a) Generate 100 observations of $Q$ where each value of $q$ is based on the simulation of drawing 50 chips from each bowl, sampling with replacement. (See part (b).) (i) Find the sample mean and sample variance of the 100 observations of $Q$. Are they close to 4 and 8, respectively? (ii) Plot a relative frequency histogram of the histogram of the observations of $Q$ with the $\chi^2(4)$ p.d.f. superimposed.

(b) Since it is known that $p_{ij} = 1/5,\ i = 1, 2, 3, 4, 5,\ j = 1, 2$, it is not necessary to estimate these probabilities. Illustrate empirically that if $(X_{i1} + X_{i2})/(n_1 + n_2)$ is replaced by $1/5$ in the definition of $Q$, then $Q^*$, say, is $\chi^2(8)$. In particular, for each set of data used for an observation of $Q$ in part (a), calculate the value of $Q^*$. (i) Find the sample mean and sample variance of the 100 observations of $Q^*$. Are they close to 8 and 16, respectively? (ii) Plot a relative frequency histogram of the observations of $Q^*$ with the $\chi^2(8)$ p.d.f. superimposed.

**9.7–2** The following table classifies 1456 people by their gender and by whether or not they favor a gun law.

| | Gender | | |
|---|---|---|---|
| Opinion | Male | Female | Totals |
| Favor | 392 | 649 | 1041 |
| Oppose | 241 | 174 | 415 |
| Totals | 633 | 823 | 1456 |

**Table 9.7–1**

Use `Contingency` to test the null hypothesis that *gender* and *opinion* are independent attributes of classification. Interpret the $p$-value of the test.

**9.7–3** A random sample of 395 people were classified by their age and by whether or not they change channels during programs when watching television. Use the following data to test the null hypothesis that *the age of the respondent* and *whether or not they change channels* are independent attributes of classification. Interpret the $p$-value of the test.

| | Age of Respondent | | | | |
|---|---|---|---|---|---|
| Change? | 18-24 | 25-34 | 35-49 | 50-64 | Totals |
| Yes | 60 | 54 | 46 | 41 | 201 |
| No | 40 | 44 | 53 | 57 | 194 |
| Totals | 100 | 98 | 99 | 98 | 395 |

**Table 9.7–2**

**9.7–4** A random sample of $n = 310$ bank customers were classified by annual household income and by where they banked, yielding the following contingency table.

| | Household Income | | | | |
|---|---|---|---|---|---|
| Location | 0-14,999 | 15,000-20,999 | 30,000-49,999 | 50,000+ | Totals |
| Main Office | 40 | 31 | 21 | 15 | 107 |
| Branch 1 | 16 | 27 | 13 | 4 | 60 |
| Branch 2 | 11 | 25 | 15 | 3 | 54 |
| Branch 3 | 6 | 18 | 9 | 3 | 36 |
| Branch 4 | 14 | 19 | 10 | 10 | 53 |
| Totals | 87 | 120 | 68 | 35 | 310 |

**Table 9.7–3**

Use the procedure `Contingency` to test the null hypothesis that *Household Income* and *Banking Location* are independent attributes. Interpret the $p$-value of the test.

**9.7–5** A random sample of bank customers was asked if they thought that the service charges were fair. The 228 customers who had an opinion were then classified by their answer and by their household income yielding the following contingency table:

| | Service Charge Fair | | |
|---|---|---|---|
| Household Income | Yes | No | Totals |
| 0-14,999 | 16 | 34 | 50 |
| 15,000-29,999 | 33 | 60 | 93 |
| 30,000-49,999 | 14 | 43 | 57 |
| 50,000+ | 4 | 24 | 28 |
| Totals | 67 | 161 | 228 |

**Table 9.7–4**

Test the null hypothesis that the two attributes are independent. Interpret the $p$-value of the test.

## Questions and Comments

**9.7–1** Use `Contingency` to work some of the exercises in your textbook.

# Chapter 10

# Nonparametric Methods

## 10.1 Order Statistics

Let $X_1, X_2, \ldots, X_n$ be a random sample of size $n$ from a distribution of the continuous type having p.d.f. $f(x)$ and distribution function $F(x)$, where $0 < F(x) < 1$ for $a < x < b$ and $F(a) = 0$, $F(b) = 0$. Let

$$\begin{aligned} Y_1 &= \text{smallest of } X_1, X_2, \ldots, X_n \\ Y_2 &= \text{second smallest of } X_1, X_2, \ldots, X_n \\ &\;\;\vdots \\ Y_n &= \text{largest of } X_1, X_2, \ldots, X_n. \end{aligned}$$

The random variables $Y_1 < Y_2 < \cdots < Y_n$ are called the order statistics of the sample. The p.d.f. of the $r$th order statistic is given by

$$g_r(y) = \frac{n!}{(r-1)!(n-r)!}[F(y)]^{r-1}[1-F(y)]^{n-r} f(y), \quad a < x < b.$$

In particular, the p.d.f.'s of the smallest and largest order statistics are, respectively,

$$g_1(y) = n[1-F(y)]^{n-1} f(y), \quad a < x < b,$$

and

$$g_n(y) = n[F(y)]^{n-1} f(y), \quad a < x < b.$$

Let $Z_r = F(Y_r)$, the cumulated probability up to and including $Y_r$. Then

$$E(Z_r) = \frac{r}{n+1}, \quad r = 1, 2, \ldots, n.$$

Furthermore,

$$E[F(Y_r) - F(Y_{r-1})] = \frac{1}{n+1}.$$

That is, the expected value of the random area between two order statistics is $1/(n+1)$.

## EXERCISES

**Purpose:** The exercises illustrate the above remarks empirically.

**10.1–1** Let $X_1, X_2, X_3$ be a random sample of size 3 from a uniform distribution on the interval (0, 1), $U(0, 10)$. Thus $f(x) = 1/10$, $0 < x < 10$, and

$$F(x) = \begin{cases} 0, & x < 0, \\ x/10, & 0 \le x < 10, \\ 1, & 10 \le x. \end{cases}$$

(a) Define $g_1(y)$, $g_2(y)$, and $g_3(y)$, the marginal p.d.f.'s of the 3 order statistics.

(b) Simulate 200 samples of size 3, say $x_1, x_2, x_3$, from the uniform distribution $U(0, 10)$. For each sample let $y_1 = \min\{x_1, x_2, x_3\}$, $y_2 = \text{median}\{x_1, x_2, x_3\}$, and $y_3 = \max\{x_1, x_2, x_3\}$.

(c) Depict a relative frequency histogram of the 200 observations of $Y_1$ along with $g_1(y)$ superimposed. Find the sample mean and sample variance of these observations. (See Exercise 10.1–2.)

(d) Depict a relative frequency histogram of the 200 observations of $Y_2$ along with $g_2(y)$ superimposed. Find the sample mean and sample variance of these observations.

(e) Depict a relative frequency histogram of the 200 observations of $Y_3$ along with $g_3(y)$ superimposed. Find the sample mean and sample variance of these observations.

**10.1–2** If $Y_1 < Y_2 < Y_3$ are the order statistics of a random sample of size $n = 3$ from the uniform distribution $U(0, \theta)$, show that

$$E(Y_r) = \frac{r\theta}{n+1}$$

and

$$\text{Var}(Y_r) = \frac{r(n-r+1)\theta^2}{(n+1)^2(n+2)}$$

for $r = 1, 2, 3$. Are the sample means and sample variances in Exercise 10.1–1 close to their respective theoretical values?

**10.1–3** Let $Y_1 < Y_2 < Y_3 < Y_4$ be the order statistics of a random sample of size 4 from the uniform distribution $U(0, 1)$. Illustrate empirically that $E(Y_r) = r/5$ and $\text{Var}(Y_r) = r(4 - r + 1)/[(4 + 1)^2(4 + 2)]$, for $r = 1, 2, 3, 4$.

**10.1–4** Let $Y_1 < Y_2 < Y_3$ be the order statistics of a random sample of size 3 from the standard normal distribution, $N(0, 1)$.

(a) Define the p.d.f.'s of $Y_1$, $Y_2$ and $Y_3$ using the procedures `NormalPDF` and `NormalCDF`.

(b) Find $E(Y_i)$ and $\text{Var}(Y_i)$, $i = 1, 2, 3$.

(c) Simulate 200 random samples of size 3 from the standard normal distribution $N(0, 1)$. For each sample of size 3 find the observed values of $Y_1$, $Y_2$, and $Y_3$.

(d) For $r = 1, 2$, and 3, graph a relative frequency histogram of the 200 observations of $Y_r$ with the respective p.d.f. superimposed.

(e) Find the sample mean and sample variance of each set of observations of $Y_1, Y_2$, and $Y_3$. Are they close to their expected values?

**10.1–5** Let $Y_1 < Y_2 < \cdots < Y_n$ be the order statistics of a random sample of size $n$ from the exponential distribution $f(x) = (1/\theta)e^{-x/\theta}$, $0 < x < \infty$. Let $1 < m < n$. Then

$$W = (2/\theta)\left[\sum_{i=1}^{m} Y_i + (n-m)Y_m\right]$$

has a chi-square distribution with $2m$ degrees of freedom. (This statistic is important in life testing experiments in which only the first $m$ "deaths" are observed.)

(a) Let $\theta = 3, n = 10, m = 4$ and illustrate empirically that the distribution of $W$ is $\chi^2(8)$. In particular, generate 50 or 100 observations of $W$. Calculate the sample mean and sample variance, comparing them with the theoretical mean and variance of $W$.

(b) Find a 90% confidence interval for $\theta$ that uses the first four order statistics and then illustrate empirically that your solution is correct. Generate 90% confidence intervals and depict the intervals using `ConfIntPlot`. You may count the number that contain $\theta = 3$ using `ConfIntSuc`.

**Remark:** See the Questions and Comments section for a way to generate observations of only the first $m$ order statistics.

**10.1–6** Let $Y_1 < Y_2 < Y_3$ be the order statistics of a random sample of size 3 from the uniform distribution $U(0, \theta)$.

(a) Prove that $Y_3$ is the maximum likelihood estimator of $\theta$.

(b) Show that $E[(4/1)Y_1] = \theta$, $E[(4/2)Y_2] = \theta$, and $E[(4/3)Y_3] = \theta$.

(c) Illustrate your answer to part (b) empirically. In particular, let $\theta = 10$. Generate 200 sets of 3 order statistics from the uniform distribution U(0,10). Show that the averages of the 200 observations of $(4/1)Y_1$, $(4/2)Y_2$, and $(4/3)Y_3$ are close to 10. Which of these statistics is the best estimator of

$\theta$? Support your answer by calculating and comparing the sample variances of the 3 sets of observations.

(d) Prove that a $100(1-\alpha)\%$ confidence interval for $\theta$ is $[y_3, y_3/\alpha^{1/3}]$.

(e) Let $\alpha = 0.20$. Generate 50 random samples of size 3 from the uniform distribution $U(0, 10)$. For each of these samples calculate the endpoints for an 80% confidence interval for $\theta$, namely $[y_3, y_3/0.20^{1/3}]$. Use `ConfIntPlot` to depict the confidence intervals. Do approximately 80% of the intervals contain $\theta = 10$? You may use `ConfIntSuc`.

**10.1–7** Let $Y_1 < Y_2 < Y_3$ be the order statistics of a random sample $X_1, X_2, X_3$ of size 3 from the uniform distribution $U(\theta - 1/2, \theta + 1/2)$. That is $Y_1 = \min\{X_1, X_2, X_3\}$, $Y_2 = \text{median}\{X_1, X_2, X_3\}$, and $Y_3 = \max\{X_1, X_2, X_3\}$. Three possible estimators of $\theta$ are the sample mean,

$$W_1 = \overline{X} = \frac{1}{3}\sum_{i=1}^{3} X_i,$$

the sample median, $W_2 = Y_2$, and the midrange, $W_3 = (Y_1 + Y_3)/2$.

(a) Simulate 100 samples of size 3 from $U(\theta - 1/2,\ \theta + 1/2)$ for a particular value of $\theta$, for example $\theta = 1/2$. For each sample, calculate the values of $W_1, W_2$, and $W_3$.

(b) By comparing the values of the sample means and sample variances of the 100 observations of each of $W_1, W_2$, and $W_3$, which of these statistics seems to be the best estimator of $\theta$?

(c) Verify that $E(W_1) = E(W_2) = E(W_3) = 1/2$ and that $\text{Var}(W_1) = 1/36$, $\text{Var}(W_2) = 1/20$, and $\text{Var}(W_3) = 1/40$ when $\theta = 1/2$. **Hint:** See Exercise 6.1-3(d).

## Questions and Comments

**10.1–1** The routine `sort` can be used for ordering sets of numbers.

**10.1–2** It is possible to simulate an experiment for which the data is collected in an ordered manner. For example, if $n$ light bulbs are turned on at the same time, the one with the shortest life burns out first, etc. Thus, there may be interest in simulating the observations of the order statistics in order, that is simulate $y_1$, then $y_2$, then $y_3$, etc.

Let $X$ be a random variable of the continuous type. The following program will generate observations of the first `m` order statistics, in order, out of a sample of size `n` where `m <= n`.

In this program, F is the distribution function of $X$ and FInv is the inverse of the distribution function of $X$. In this particular example, we are sampling from an exponential distribution with mean $\theta = 3$.

```
n := 10;
m := 4;
F := x -> 1/3*exp(-1/3*x);
FInv := y -> -3*ln(1-y);
for i from 1 to 100 do
YY[i][1] := FInv(1-rng()^(1/n));
for j from 1 to m-1 do
YY[i][j+1] := FInv(1-(1-evalf(F(YY[i][j])))*
(1-rng())^(1/(n-j)))
od
od:
Y1 := [seq(YY[k][1],k = 1 .. 100)];
Y2 := [seq(YY[k][2],k = 1 .. 100)];
Y3 := [seq(YY[k][3],k = 1 .. 100)];
Y4 := [seq(YY[k][4],k = 1 .. 100)];
```

The above procedure is based on the following theoretical development:

Let $Y_1 < Y_2 < \cdots < Y_n$ be the order statistics of a random sample of size $n$ from a continuous distribution that has a distribution function $F(x)$, where $0 < F(x) < 1$ for $a < x < b$, $F(a) = 0$, $F(b) = 1$. The marginal p.d.f. of $Y_i$ is

$$g_i(y_i) = \frac{n!}{(n-i)!(i-1)!}[1 - F(y_i)]^{n-i}[F(y_i)]^{i-1}f(y_i), \quad a < y_i < b.$$

The joint marginal p.d.f. of $Y_i$ and $Y_{i+1}$ is

$$g_{i,i+1}(y_i, y_{i+1}) = \frac{n!}{(i-1)!(n-i-1)!}[F(y_i)]^{i-1}[1 - F(y_{i+1}]^{n-i-1}f(y_i)f(y_{i+1}),$$

for $a < y_i < y_{i+1} < b$. The conditional p.d.f. of $Y_{i+1}$, given $Y_i = y_i$, is defined by

$$\begin{aligned} h(y_{i+1}|y_i) &= \frac{g_{i,i+1}(y_i, y_{i+1})}{g_i(y_i)} \\ &= \frac{(n-i)[1 - F(y_{i+1})]^{n-i-1}f(y_{i+1})}{[1 - f(y_i)]^{n-1}}, \quad y_i < y_{i+1} < b. \end{aligned}$$

The conditional distribution function of $Y_{i+1}$, given $Y_i = y_i$, is defined by

$$H(y_{i+1}|y_i) = \int_{y_i}^{y_{i+1}} h(x|y_i)\, dx$$

$$= 1 - \left[\frac{1 - F(y_{i+1})}{1 - F(y_i)}\right]^{n-1}, \quad y_i < y_{i+1} < b.$$

The distribution function of $Y_1$, the first order statistic, is

$$v = G_1(y_1) = 1 - [1 - F(y_1)]^n.$$

Solving for $y_1$ we obtain

$$y_1 = F^{-1}(1 - [1 - v]^{1/n}).$$

Let $V$ be $U(0,1)$. Then $1 - V$ is also $U(0,1)$. By the method given in Section 4.7, we can generate the value of the first order statistic.

Now let

$$v = H(y_{i+1}|y_i) = 1 - \left[\frac{1 - F(y_{i+1})}{1 - F(y_i)}\right]^{n-1}.$$

Then

$$\begin{aligned} 1 - F(y_{i+1}) &= [1 - F(y_i)](1 - v)^{1/(n-i)} \\ y_{i+1} &= F^{-1}[1 - \{1 - F(y_i)\}\{1 - v\}^{1/(n-i)}]. \end{aligned}$$

Thus, given the value of $y_i$, the observed value of $Y_{i+1}$ can be found, again using the method given in Section 4.7.

**Remark**: The above discussion is based on the article "Machine-Generation of Order Statistics for Monte Carlo Computations," *The American Statistician*, February, 1972, pages 26-27, by D. Lurie and H.O. Hartly.

**10.1–3** Work Exercise 10.1–5 using the above method and generating only the observed values of the first 4 of the 10 order statistics.

**10.1–4** There may be times when we are interested in simulating observations of the order statistics, in order, beginning with the largest. For example, this could be true in an admittance procedure for a medical school or a prestigious university. The largest `m` out of a random sample of `n` observations of $X$ can be generated using the following program. The example used in this procedure is that of generating order statistics from a $U(0, 10)$ distribution, with $n = 9$. Thus $E(Y_i) = i/10$ and this can be checked empirically. In this program, `F` is the distribution function of $X$ and `FInv` is the inverse of the distribution function of $X$.

```
n := 9;
m := 3;
F := x -> x/10;
FInv := y -> 10*y;
for k from 1 to 200 do
YY[k][n] := FInv(rng()^(1/n));
for i from n-1 by -1 to n-m+1 do
YY[k][i] := FInv(F(YY[k][i+1]*rng()^(1/i)))
od
od:
for i from n by -1 to n-m+1 do
Y[i] := [seq(YY[k][i],k = 1 .. 200)]
od:
seq(Mean(Y[j]),j = 7 .. 9);
```

**Remark**: The above program is based on the material in the note "Machine Generation of Order Statistics," *The American Statistician*, October 1972, pages 56-57, by Hugh A. Reeder.

## 10.2 Confidence Intervals for Percentiles

Let $Y_1 < Y_2 < \cdots < Y_n$ be the order statistics of a random sample of size $n$ from a continuous type distribution. Let $m$ denote the median of this distribution. Select $i$ and $j$ so that

$$P(Y_i < m < Y_j) = \sum_{k=i}^{j-1} \binom{n}{k} \left(\frac{1}{2}\right)^k \left(\frac{1}{2}\right)^{n-k} = 1 - \alpha.$$

Then the observed interval $(y_i, y_j)$ is a $100(1-\alpha)\%$ confidence interval for $m$.

### EXERCISES

**Purpose:** The exercises illustrate confidence intervals for medians.

**10.2–1** Let $X$ be $U(0,1)$.

(a) Simulate 50 random samples of size $n = 11$ from this uniform distribution. For each sample find the values of $Y_3$ and $Y_9$, the third and ninth order statistics. Store these values in a list of lists. **Hint**: You may use the routine `sort` for ordering each sample.

(b) Verify that $(y_3, y_9)$ is a 93.46% confidence interval for the median, $m = 0.5$, of the uniform distribution $U(0,1)$.

(c) Approximately 93.46% of the confidence intervals generated in part (a) should contain the median. To illustrate this, use `ConfIntPlot` to depict the confidence intervals.

(d) What proportion of your confidence intervals contain the median? You may use `ConfIntSuc`.

**10.2–2** Let $X$ be $N(16.4, 2.9)$.

(a) Simulate 50 random samples of size $n = 16$ from this distribution. For each sample, find the values of $Y_5$ and $Y_{12}$, the 5th and 12th order statistics. Store these values in a list of lists. **Hint:** You may use the routine `sort`.

(b) Verify that $(y_5, y_{12})$ is a 92.32% confidence interval for $m = 16.4 = \mu$.

(c) Approximately 92.32% of the 50 confidence intervals from part (a) should contain the median. To illustrate this, use the subroutine `ConfIntPlot` to depict the confidence intervals.

(d) What proportion of the confidence intervals contain the median?

(e) How do the lengths of these confidence intervals compare with those based on the sample mean? You may use `ConfIntAvLen` to help answer this question.

**10.2–3** Let $X$ have an exponential distribution with mean $\theta = 10$.

(a) What is the value of the median $m$?

(b) Generate 50 random samples of size 8 from this distribution. For each sample find the values of $Y_2$ and $Y_7$, the 2nd and 7th order statistics. Store these values in a list of lists.

(c) Verify that $(y_2, y_7)$ is a 92.97% confidence interval for $m$.

(d) Use `ConfIntPlot` to depict the confidence intervals. Do approximately 92.96% of the confidence intervals contain the median?

(e) Find $\overline{y}_2$ and $\overline{y}_7$, the sample means of the observations of $Y_2$ and $Y_7$, respectively. Are they close to the theoretical values, $E(Y_2)$ and $E(Y_7)$, respectively?

## Questions and Comments

**10.2–1** In Exercise 10.2–1, for the fifty 93.46% confidence intervals, let $Y$ equal the number of intervals that contain the median $m$. How is $Y$ distributed?

**10.2–2** In Exercise 10.2–1, what proportion of the confidence intervals generated by the entire class contain the median?

## 10.3 Binomial Tests for Percentiles

Let $m$ be the median of the distribution of $X$. We shall test the hypothesis $H_0: m = m_0$. Given a random sample $X_1, X_2, \ldots, X_n$, let $Y$ equal the number of negative signs among

$$X_1 - m_0, X_2 - m_0, \ldots, X_n - m_0.$$

If $H_0$ is true, the distribution of $Y$ is $b(n, 1/2)$. If the alternative hypothesis is $H_1: m > m_0$, we would reject $H_0$ if the observed value of $Y$ is too small, say $y < c_1 < n/2$. And if the alternative hypothesis is $H_1: m < m_0$, we would reject $H_0$ if the observed value of $Y$ is too large, say $y > c_2 > n/2$. The constants $c_1$ and $c_2$ are selected to yield the desired significance level. This test is often called the sign test.

### EXERCISES

**Purpose:** The sign test is illustrated. A comparison of the sign test with a test using the sample mean is given.

**10.3–1** We shall use the sign test to test the hypothesis $H_0: m = 6.2$ against the alternative hypothesis $H_1: m < 6.2$, where $m$ is the median of the distribution. The test will be based on a random sample of size $n = 20$.

(a) Let Y equal the number of observations in a random sample of size 20 that are less than 6.2. Show that $C = \{y : y > 13\} = \{y : y \geq 14\}$ is a critical region of size $\alpha = 0.0577$.

(b) Suppose that $X$ has an exponential distribution with a mean $\mu = \theta$. Show that if $m = 6.2$ is the median, then $\theta = 8.945$ is the mean. Furthermore, show that in general, $\theta = -m/\ln(0.5) = m/\ln 2$.

(c) For $m = 3.5, 3.8, \ldots, 6.2$, generate 50 random samples of size 20 from an exponential distribution with a mean of $\theta = m/\ln 2$. For each value of $m$, count the number of times that $y > 13$ and thus $H_0$ was rejected.

(d) Plot the empirically defined power function using your generated data.

**10.3–2** Let $Z_1$ and $Z_2$ have independent normal distributions $N(0, 1)$. Let $X = Z_1/Z_2$. Then $X$ has a Cauchy distribution with p.d.f. $f(x) = 1/[\pi(1 + x^2)]$, $-\infty < x < \infty$. The median of this Cauchy distribution is $m = 0$.

(a) Show that $\mu = E(X)$ does not exist.

(b) We shall test $H_0: m = 0$ against $H_1: m \neq 0$. Let $Y$ equal the number of observations that are less than 0 in a sample of size 14 from a Cauchy distribution. Let $C = \{y : y \leq 3 \text{ or } y \geq 11\}$. Show that $\alpha = 0.0574$ is the significance level of this test.

(c) Simulate 50 samples of size 14 from the Cauchy distribution. For each sample calculate the value of $y$. Print the 14 observations and the value of

$y$ on one or two lines for each of the 50 repetitions. How do the sample values from the Cauchy distribution compare with samples from a normal distribution $N(0,1)$?

(d) What proportion of the 50 samples in part (c) led to rejection of $H_0$? Is this proportion close to $\alpha$?

(e) Could this test be based on $\overline{X}$? If the answer is yes, define a critical region that has significance level $\alpha = 0.0574$. For each sample in part (c), determine whether $H_0$ was rejected. Was $H_0$ rejected about $\alpha$ of the time? Why?

## Questions and Comments

**10.3–1** Could you work Exercise 10.3–1 basing your test on $\overline{X}$?

**10.3–2** In the exercises, since we were simulating samples, we knew the underlying distribution. Remember that the sign test is distribution free and, in applications, the underlying distribution is usually not known.

# 10.4 The Wilcoxon Test

The Wilcoxon test can be used to test the hypothesis $H_0: m = m_0$ against an alternative hypothesis. Given a random sample $X_1, X_2, \ldots, X_n$ from a continuous distribution which is symmetric about its median, rank the absolute values of the differences, $|X_1 - m_0|, \ldots, |X_n - m_0|$, in ascending order. Let $R_i$ denote the rank of $|X_i - m_0|$. With each $R_i$ associate the sign of the difference $X_i - m_0$. Let $W$ equal the sum of these signed ranks. If $n$ is sufficiently large,

$$Z = \frac{W - 0}{\sqrt{n(n+1)(2n+1)/6}}$$

is approximately $N(0,1)$ when $H_0$ is true.

## EXERCISES

**Purpose:** The exercises illustrate the Wilcoxon test and compare it with the sign test and a test based on the sample mean.

**10.4–1** Let $X$ be $N(\mu, 6)$. We shall test the hypothesis $H_0: m = 6.2$ against an alternative hypothesis $H_1: m < 6.2$ using (i) the sign test, (ii) the Wilcoxon test, and (iii) the $z$–test or $t$–test based on $\overline{x}$. Note that for the normal distribution, the median, $m$, and the mean, $\mu$, are equal. We shall use a random sample of size $n = 20$.

(a) Let $Y$ equal the number of observations that are less than 6.2. If the critical region is $C = \{y : y \geq 14\}$, what is the significance level, $\alpha$, of this test?

(b) Let $W$ equal the sum of the signed ranks, the Wilcoxon statistic. If the critical region is $C = \{w : w \leq -88.1266\}$, what is the approximate value of the significance level, $\alpha$, of this test?

(c) Let $\overline{X}$ be the sample mean. If the critical region is $C = \{\overline{x} : \overline{x} \leq 5.299\}$, what is the significance level, $\alpha$, of this test? Use either a $z$ or $t$ statistic.

(d) For $m = 4.4, 4.6, \ldots, 6.2$, generate 30 samples of size 20 from the normal distribution $N(m, 6)$. For each value of $m$, (i) calculate and print on the screen 30 values of $y, w$, and $\overline{x}$, (ii) count the numbers of times that $y \geq 14$, $w \leq -88.1266$, and $\overline{x} \leq 5.299$ and thus $H_0$ was rejected.

(e) Plot the empirically defined power function using the values of $Y$. If possible, add the theoretically defined power function.

(f) Plot the empirically defined power function using the values of $W$.

(g) Plot the empirically defined power function using the values of $\overline{X}$. If possible, add the theoretically defined power function.

(h) Comment on the comparison of the three power functions. Which is the most powerful test?

**10.4–2** Describe a method for determining when $n$ is sufficiently large so that $Z = W/\sqrt{n(n+1)(2n+1)/6}$ is approximately $N(0, 1)$.

## Questions and Comments

**10.4–1** Suppose that the underlying distribution in Exercise 10.4–1 had not been a normal distribution. What effect would that have on the comparison of the three power functions?

**10.4–2** The procedure `Wilcoxon` can be used to find the value of the Wilcoxon statistic.

# 10.5 Two–Sample Distribution–Free Tests

Let $X$ and $Y$ be independent random variables of the continuous type with respective medians $m_X$ and $m_Y$ and distribution functions $F(x)$ and $G(y)$. To test the hypothesis $H_0 : m_X = m_Y$, take random samples $X_1, \ldots, X_{n_1}$ and $Y_1, \ldots, Y_{n_2}$ from the two distributions.

For the median test, let $V$ equal the number of $X$ values in the lower half of the combined sample. If the alternative hypothesis is $H_1 : m_X < m_Y$, reject $H_0$ if $V$ is too large. If the alternative hypothesis is $H_1: m_X > m_Y$, reject $H_0$ if $V$ is too small. When $F(z) = G(z)$ and $n_1 + n_2 = 2k$, where $k$ is a positive integer, $V$ has a hypergeometric distribution with p.d.f.

$$h(v) = P(V = v) = \frac{\binom{n_1}{v}\binom{n_2}{k-v}}{\binom{n_1+n_2}{k}}, \quad v = 0, 1, 2, \ldots, n_1.$$

For the Wilcoxon test, order the combined sample in increasing order of magnitude. Assign to the ordered values the ranks $1, 2, \ldots, n_1+n_2$. Let $W$ equal the sum of the ranks of $Y_1, Y_2, \ldots, Y_{n_2}$. The critical region for testing $H_0: m_X = m_Y$ against $H_1: m_X < m_Y$ is of the form $W \geq c$. If the alternative hypothesis is $H_1: m_X > m_Y$, the critical region is of the form $W \leq c$. If $F(z) = G(z)$, $n_1 > 7$, and $n_2 > 7$, then

$$\begin{aligned} \mu_W &= n_2(n_1 + n_2 + 1)/2, \\ \sigma_W^2 &= n_1 n_2(n_1 + n_2 + 1)/12 \end{aligned}$$

and

$$Z = (W - \mu_W)/\sigma_W$$

is approximately $N(0, 1)$.

## EXERCISES

**Purpose:** The powers of the median test, the Wilcoxon test, and the $t$ test are compared. The normality of $(W - \mu_W)/\sigma_W$ is investigated.

**10.5–1** Let $X$ be $N(\mu_X, 24)$ and let $Y$ be $N(\mu_Y, 24)$. Recall that for the normal distribution, the median and mean are equal. Consider a test of the hypothesis $H_0: m_X = m_Y$ against the alternative hypothesis $H_1: m_X > m_Y$ based on random samples of sizes $n_1 = 8$ and $n_2 = 10$. In this exercise we shall compare the powers of the median test, the Wilcoxon test, and the $t$-test. Let $V$ equal the number of $X$ values out of the 8 observations of $X$ that are in the lower half of the combined sample. Let $W$ equal the sum of the ranks of $Y_1, \ldots, Y_{10}$ when the combined sample has been ordered in increasing order of magnitude.

(a) (i) Show that $C = \{v : v \leq 2\}$ is a critical region of size $\alpha = 0.0767$ when the median test is used. (ii) If the Wilcoxon test is used, find a critical region yielding an approximate significance level of $\alpha = 0.0767$ using the normal approximation. (iii) Since we are sampling from independent normal distributions, the $t$ test statistic with 16 degrees of freedom is appropriate. Use TP to determine a critical region of size $\alpha = 0.0767$.

(b) Let $\mu_X = m_X = 110$. For each value of $\mu_Y = 110, 109, 108, \ldots, 101$, generate 30 observations values of (i) $V$, the median test statistic, (ii) $W$, the Wilcoxon test statistic, and (iii) the $T$, the $t$ statistic. (Note that each of these uses a random sample of size 8 from the normal distribution, $N(110, 24)$ and a random sample of size 10 from the normal distribution, $N(\mu_Y, 24)$.) And for each value of $\mu_Y$ and each test, count the numbers of times that $H_0$ was rejected.

(c) Plot the empirically defined power functions based on the observations of (i) $V$, (ii) $W$, and (iii) $T$.

(d) Comment on the comparison of the three power functions.

**10.5–2** Let $X$ be $N(110, 100)$, let $Y$ be $N(110, 100)$, and let $X$ and $Y$ be independent.

(a) Let $n_1 = 8$ and $n_2 = 10$. Generate 100 observations of the Wilcoxon statistic, $W$.

(b) Calculate the sample mean and sample variance of your observations and compare them with the theoretical value.

(c) Let $Z = (W - \mu_W)/\sigma_W$. Plot a relative frequency histogram of your 100 transformed observations of $W$ with the $N(0, 1)$ p.d.f. superimposed.

(d) Find the value of the chi-square goodness of fit statistic.

(e) Use `KSFit` to plot the empirical distribution function of the observations of $Z$ with the $N(0, 1)$ distribution function superimposed.

(f) On the basis of your data, does $Z$ seem to be $N(0, 1)$?

**10.5–3** Repeat Exercise 10.5–2 with other values of $n_1$ and $n_2$. Try to determine the values of $n_1$ and $n_2$ for which $Z$ is $N(0, 1)$.

### Questions and Comments

**10.5–1** In the exercise we sampled from normal distributions. In applications the assumption of normality is not made.

**10.5–2** The subroutine `Wilcoxon2` can be used to find values of the two-sample Wilcoxon statistic.

## 10.6 Run Test and Test for Randomness

Let $X$ and $Y$ be random variables of the continuous type with distribution functions $F(x)$ and $G(y)$, respectively. Given $n_1$ observations of $X$ and $n_2$ observations of $Y$, let $R$ equal the number of runs when the observations are placed into one collection in ascending order. A test of $H_0\colon F(z) = G(z)$ can be based on the number of runs. The hypothesis is rejected if the number of runs is too small.

When $H_0$ is true, the p.d.f. of $R$ is given by

$$P(R = 2k) = 2 \cdot \frac{\binom{n_1-1}{k-1}\binom{n_2-1}{k-1}}{\binom{n_1+n_2}{n_1}}$$

and

$$P(R = 2k+1) = \frac{\binom{n_1-1}{k}\binom{n_2-1}{k-1} + \binom{n_1-1}{k-1}\binom{n_2-1}{k}}{\binom{n_1+n_2}{n_1}}$$

for $2k$ and $2k+1$ in the space of $R$.

When $n_1$ and $n_2$ are large, say each is greater than 10,

$$Z = \frac{R - \mu_R}{\sqrt{\text{Var}(R)}}$$

is approximately $N(0,1)$, where

$$\begin{aligned} \mu_R &= \frac{2n_1n_2}{n_1+n_2} + 1, \\ \text{Var}(R) &= \frac{(\mu_R - 1)(\mu_R - 2)}{n_1 + n_2 - 1} \\ &= \frac{2n_1n_2(2n_1n_2 - n_1 - n_2)}{(n_1+n_2)^2(n_1+n_2-1)}. \end{aligned}$$

## EXERCISES

**Purpose:** Applications of the run test are given and they are compared with tests based upon normal sampling theory. The approximate normality of $R$ is investigated.

**10.6–1** Let $X$ be $N(\mu_X, 24)$ with distribution function $F(x)$ and let $Y$ be $N(\mu_Y, 24)$ with distribution function $G(y)$. We shall test $H_0: F(z) = G(z)$ based on random samples of sizes $n_1 = 8$ and $n_2 = 10$, respectively. Let $R$ denote the number of runs in the combined ordered sample.

(a) Show that $C = \{r : r \le 7\}$ is a critical region of size $\alpha = 0.117$.

(b) Let $\mu_X = m_X = 110$. For each of $\mu_Y = 110, 109, \ldots, 101$, generate 30 values of $R$. Note that each value of $R$ depends on a random sample of size 8 from the normal distribution, $N(\mu_Y, 100)$ and a random sample of size 10 from the normal distribution, $N(\mu_Y, 100)$. For each value of $\mu_Y$, count the number of times that $r \in C$ and thus $H_0$ was rejected.

(c) Plot the empirically defined power function.

(d) Comment on the comparison of this power function with those generated in Exercise 10.5–1.

**10.6–2** Let $X$ be $N(0, \sigma_X^2)$ with distribution function $F(x)$ and let $Y$ be $N(0, \sigma_Y^2)$ with distribution function $G(y)$. We shall test $H_0: F(z) = G(z)$ based on random samples of sizes $n_1 = 8$ and $n_2 = 10$, respectively. Let $R$ denote the number of runs in the combined sample.

(a) Show that $C = \{r : r \leq 7\}$ is a critical region of size $\alpha = 0.117$.

(b) Let $\sigma_Y^2 = 1$. Let $\sigma_X^2 = 1, 2, \ldots, 9$ and generate 30 values of $R$. Note that each value of $R$ depends on a random sample of size 8 from $N(0, \sigma_X^2)$ and a random sample of size 10 from $N(0, 1)$. For each value of $\theta$, count the number of times that $r \in C$ and thus $H_0$ was rejected. (See part (e).)

(c) Plot the empirically defined power function.

(d) Since we are sampling from independent normal distributions, the $F$ test statistic is appropriate. Use `FP` to determine a critical region of size $\alpha = 0.117$.

(e) For each $\sigma_X^2$ and each pair of samples in part (b), find the value of the $F$ statistic. Count the number of times that $H_0$ was rejected using the $F$ statistic.

(f) Plot the empirically defined power function. You could also add the theoretically defined power function.

(g) Comment on the comparison of the power functions in parts (c) and (f).

**10.6–3** Determine empirically the values of $n_1$ and $n_2$ for which $R$ has an approximate normal distribution.

**10.6–4** Use a run test to test the random number generator, `RNG`, for randomness. Base your test on a random sample of $k = 20$ random numbers.

## Questions and Comments

**10.6–1** In the exercises we sampled from normal distributions and thus could compare the power function using the run test with the power function based on normal sampling theory. In applications the only assumption about the underlying distributions is that they are continuous.

**10.6–2** Is your answer to Exercise 10.6–3 consistent with the rule of thumb that $n_1$ and $n_2$ should both be larger than 10?

**10.6–3** The procedure `Runs` can be used to determine the number of runs.

# 10.7 Kolmogorov–Smirnov Goodness of Fit Test

The Kolmogorov–Smirnov goodness of fit test is based on the closeness of the empirical and hypothesized distribution functions. Consider a test of the hypothesis $H_0: F(x) =$

$F_0(x)$, where $F_0(x)$ is the hypothesized distribution function for the random variable $X$. Let $F_n(x)$ be the empirical distribution function based on a random sample of size $n$. The Kolmogorov-Smirnov statistic is defined by

$$D_n = \sup_x[|F_n(x) - F_0(x)|].$$

The hypothesis $H_0$ is rejected if $D_n$ is too large.

## EXERCISES

**Purpose:** The Kolmogorov–Smirnov statistic is used to test some simulations from various distributions. The distribution of the Kolmogorov–Smirnov statistic is illustrated empirically.

**10.7–1** (a) Generate 100 samples of 10 random numbers. For each sample calculate the value of the Kolmogorov–Smirnov statistic with $F_0(x) = x$, $0 < x < 1$.

(b) Plot a relative frequency histogram of your data.

(c) For your data are approximately 20% of your observations of $D_{10}$ greater than 0.32?

**10.7–2** (a) Simulate 50 observations from an exponential distribution with a mean of $\theta = 2$.

(b) Use `KSFit` to plot the theoretical and empirical distribution functions. Note that this procedure gives the value of the Kolmogorov–Smirnov statistic.

(c) At a significance level of $\alpha = 0.10$, do you accept the hypothesis that the sample comes from an exponential distribution with a mean of $\theta = 2$?

**10.7–3** Repeat Exercise 10.7–2 simulating 50 observations from a chi–square distribution with $r = 6$ degrees of freedom. Make the necessary changes in the questions.

**10.7–4** Repeat Exercise 10.7–2 simulating observations from the standard normal distribution, $N(0, 1)$. Make the necessary changes in the questions.

**10.7–5** A chemistry laboratory experiment was conducted to determine the concentration of $CaCO_3$ in mg/l. A random sample was taken from the same source by each of 30 students resulting in the data in Table 10.7–1.

| | | | | | |
|---|---|---|---|---|---|
| 129.9 | 131.5 | 131.2 | 129.5 | 128.0 | 131.8 |
| 127.7 | 132.7 | 131.5 | 127.8 | 131.4 | 128.5 |
| 134.8 | 133.7 | 130.8 | 130.5 | 131.2 | 131.1 |
| 131.3 | 131.7 | 133.9 | 129.8 | 128.3 | 130.8 |
| 132.2 | 131.4 | 128.8 | 132.7 | 132.8 | 131.2 |

**Table 10.7–1**

(a) Calculate the sample mean and sample variance of these data.

(b) Test the hypothesis that these data are values of a random sample from a normal distribution. Use the sample mean and sample variance for $\mu$ and $\sigma^2$, respectively. Plot the data using `KSFit` and `NormalCDF` and base your decision on the Kolmogorov–Smirnov goodness of fit statistic.

**10.7–6** Many textbook exercises make assumptions about the distribution from which a sample is taken. Check several of these in your book to determine whether the assumption is correct.

## Questions and Comments

**10.7–1** Did the subroutines that simulate random samples from various distributions perform satisfactorily? Compare your conclusions with those of other members of the class.

# Appendix A

# A Brief Introduction to Maple

Much of what is done in high school and college mathematics involves the manipulation of symbols or expressions composed of symbols. For example, the equation

$$5x + 4 = 0$$

is solved (for $x$) by manipulating the symbols $x$, $+$ and $=$ and the numbers 5, 4 and 0, which themselves can be interpreted symbolically. The solution of

$$x^2 + x - 2 = 0$$

is essentially all symbolic with

$$0 = x^2 + x - 2 = (x - 1)(x + 2)$$

leading to the solutions $x = 1$ and $x = -2$.
While many such examples are easy to do "by hand," others such as the simplification of

$$\frac{\sin^2 x + \sin^3 x \cos x + \sin x \cos^3 x - \sin x \cos x}{\sin x \cos x}$$

to $\tan x$ are more cumbersome.

*Maple* is an easy to use system that performs symbolic, numeric, and graphic computations. It is available on many modern computers, including IBM-PC compatible microcomputers (80386 or 80486 based systems, for both DOS and Microsoft Windows environments) and Macintosh computers. We will assume here that the reader is able to access *Maple* on some type of computer.

To enhance *Maple*'s applicability to problems in probability and statistics, we have supplemented *Maple* with about 130 special procedures. This Appendix describes some of the most basic features of *Maple*; Appendix B deals with the statistical supplement to *Maple* that is provided with this book.

## A.1 Basic Syntax

Users first **connect** to *Maple*, then they **interact** with *Maple* through a sequence of commands and, when done, they **disconnect** from *Maple*. A *Maple* session consists of a sequence of commands when *Maple* displays its > prompt. In the following examples the user does not type >, *Maple* provides this prompt. There are several things you need to remember as you work with *Maple*.

1. Each *Maple* command should terminate with a ";" and the return key (on Macintosh computers this could be the enter key). When a command is issued, *Maple* provides a response and a new prompt. (Actually, any number of commands may be placed on a line, but they are not executed until the return key is pressed.

2. It is easy to "confuse" *Maple* by typing incorrectly formed commands, making it difficult to interpret subsequent correctly formed commands. This problem can be avoided by typing a ";" following an error.

3. *Maple* is **case sensitive** (i.e., lower and upper case letters cannot be used interchangeably.) Almost all *Maple* commands and functions are in lower case. Two notable exceptions are "Pi" and "E" which stand for $\pi$ and $e$, respectively.

## A.2 Maple Expressions

Analogous to usual mathematical notation, simple expressions in *Maple* are constructed from variables, constants and operators. For the moment suppose that +, −, * (multiplication), /, and ^ (exponentiation—this key is shift-6 on the keyboard) are the only operations. These five operations are grouped into three **precedence levels**

| | |
|---|---|
| ^ | highest precedence |
| *, / | |
| +, − | lowest precedence |

When an expression is evaluated, operations are performed (from left to right) in the order of their precedence (from highest to lowest). Thus,

```
> 7 + 4 * 3;
```

would produce 19 and not 33. If your intent is to perform the addition before the multiplication, then you should enter

```
> (7 + 4) * 3;
```

In many situations, it is customary in mathematics to indicate a multiplication **implicitly**. For example, it is understood that

$$(x - y)(x + y)$$

implicitly mandates the multiplication of $x - y$ and $x + y$. **All multiplication in *Maple* must be explicit**. The product mentioned above will have to be entered as

```
> (x - y) * (x + y);
                        (x - y)(x + y)
```

Notice that you have to make the expression understandable to *Maple* by including the "*". Maple returns the compliment by making its output convenient to you by omitting the "*".

When two or more operations of the same precedence appear in an expression, these operations are performed in left-to-right order during the evaluation of the expression. To form complex expressions, parentheses are used to group subexpressions. Consistent with common practice, the deepest nested subexpression is evaluated first.

Given the *Maple* expression

```
> y + (x * (a - b)^3)/b;
```

in what order will these operations be performed during its evaluation? (Ans: –, ^, *, /, +) Suppose further that $a = 3$, $b = 5$, $x = 10$, and $y = 2$. What is the value of the expression? (Ans: –14)

## A.3 Assignments and Recall

For convenience, *Maple* allows users to make assignments (i.e., give names to numbers, expressions,etc.). Instead of simply entering the expression $x^2 + 2x + 1$ into *Maple* by

```
> x^2 + 2 * x + 1;
                          2
                         x  + 2 x + 1
```

you can choose to give this expression the name $a$ with

```
> a:= x^2 + 2 * x + 1;
                              2
                         a:= x  + 2 x + 1
```

Later on, you can recall this saved expression by

```
> a;
                          2
                         x  + 2 x + 1
```

The name $a$ can also be used in subsequent work as in

```
> b:= (x + 1) * a^2;
                                     2           2
                          (x + 1) (x  + 2 x + 1)
```

Suppose that you now wish to obtain $a^2 + b^3$. You can do this by

```
> a^2 + b^3;
                     2            2               3    2             6
                   (x  + 2 x + 1)  + (x + 1)  (x  + 2 x + 1)
```

If you decide to save this last expression, you can give it name $c$ with

```
> c:= ";
                     2            2               3    2             6
              c:= (x  + 2 x + 1)  + (x + 1)  (x  + 2 x + 1)
```

The " (double quote—located near the shift key on the lower right side of your terminal) in *Maple* refers to the result of the immediately preceding command, disregarding any error messages. "" (two double quotes) refers to the output before the last one, and """ (three double quotes) refers to the output before that—again disregarding error messages. If you wish, you can reassign a new expression to $a$ at any time as with

```
> a:=c/b;
                  2            2               3    2             6
                (x  + 2 x + 1)  + (x + 1)  (x  + 2 x + 1)
         a:= -------------------------------------------
                                  2           2
                           (x + 1) (x  + 2 x + 1)
```

## A.4 Maple Notation

Two unusual notational features of *Maple*: the explicit use of $*$ (the multiplication operator) and the use of ", "", """ have already been mentioned. *Maple* also departs from customary mathematical notation when certain internally defined functions, such as trigonometric functions, are used. In common mathematical notation, $\sin x$ designates the sine function applied to $x$ (not sin times $x$). This intent would be clearer if the explicit functional notation $\sin(x)$ were used instead. This is precisely what must be done in *Maple*.

The use of exponents in trigonometric expressions may be even more confusing. In *Maple*, the familiar identity $\sin^2 x + \cos^2 x = 1$ must be entered as

```
> sin(x)^2 + cos(x)^2 = 1;
                               2          2
                         sin(x)  + cos(x)  = 1
```

## A.5 A Sample Session

If you do not have prior experience with *Maple*, the following *Maple* session will introduce you to some of the most commonly used *Maple* commands. We recommend that you enter the commands that are described below and compare the results that you get with those included as part of our interaction with *Maple*.

1. Connect to *Maple*

2. Enter and simplify the expression $(1 + 1/n + 1/m)(1 + 1/n - 1/m)^{-1}$. One way of doing this is to enter the entire expression in one step:

```
> (1 + 1/n + 1/m) * (1 + 1/n - 1/m)^(-1);
                                   1 + 1/n + 1/m
                                   -------------
                                   1 + 1/n - 1/m
```

   followed by

```
> simplify (");
                                    nm + m + n
                                    ----------
                                    nm + m - n
```

3. Inexperienced users are likely to make frequent typing errors. Therefore, the following, seemingly more complicated way of arriving at the desired simplification, is recommended.

```
> a:= 1 + 1/n + 1/m;
                                a: = 1 + 1/n + 1/m
>b: =1 + 1/n - 1/m;
                                b: = 1 + 1/n - 1/m
> a / b;
                                   1 + 1/n + 1/m
                                   -------------
                                   1 + 1/n - 1/m
> simplify (");
                                    nm + m + n
                                    ----------
                                    nm + m - n
```

4. Enter and simplify the expression

$$\frac{\sin^2 x + \sin^3 x \cos x + \sin x \cos^3 x - \sin x \cos x}{\sin x \cos x}$$

   This can be done through the following sequence of *Maple* commands.

```
> a:= sin(x)^2 + sin(x)^3 * cos(x);
> b:= sin(x) * cos(x)^3 - sin(x) * cos(x);
> c:= sin(x) * cos(x);
> (a + b)/c;
> simplify (");
```

5. Simple arithmetic operations can be performed as follows.

```
> 3*5+7;
                                        22
```

6. $\pi$ is shown symbolically (that is the only way it can expressed in exact form). However, we can obtain numerical approximations to $\pi$ to any level of accuracy.

```
> Pi;
                                        Pi

> evalf(Pi);
                                   3.141592654

> evalf(Pi, 1000);
3.14159265358979323846264338327950288419716939937510582097494459230781640628620\
0899862803482534211706798214808651328230664709384460955058223172535940812848111\
1745028410270193852110555964462294895493038196442881097566593344612847564823378\
8678316527120190914564856692346034861045432664821339360726024914127372458700660\
0631558817488152092096282925409171536436789259036001133053054882046652138414695\
5194151160943305727036575959195309218611738193261179310511854807446237996274956\
6735188575272489122793818301194912983367336244065664308602139494639522473719070\
0217986094370277053921717629317675238467481846766940513200056812714526356082778\
8577134275778960917363717872146844090122495343014654958537105079227968925892354\
4201995611212902196086403441815981362977477130996051870721134999999837297804995\
5105973173281609631859502445945534690830264252230825334468503526193118817101000\
0313783875288658753320838142061717766914730359825349042875546873115956286388235\
5378759375195778185778053217122680661300192787661119590921642019
```

```
> 100!;
9332621544394415268169923885626670049071596826438162146859296389521759999322991\
5608941463976156518286253697920827223758251185210916864000000000000000000000000\
00
```

7. Algebraic equations can be solved and expressions can be manipulated.

```
> p:=-6*x^2 - 6*x + 72;
```

```
                                      2
                           p := - 6 x  - 6 x + 72
> solve(p,x);
                                    -4, 3
> subs(x=5,p);
                                    -108

> x:=5; p;
                                    x:=5
                                    -108

> x:='x';p;
                                    x:=x
                                  2
                             - 6 x  - 6 x + 72
```

8. If necessary, use the *Maple* help system to explain what each of the following steps does. To obtain help on a specific *Maple* command, you type ? followed by the command name (without any arguments). For example, to see an explanation and examples of how to use the the `solve` command, you simply type `?solve`.

```
> f:=(5*x^2+2*x-8)^5;
                                      2           5
                            f := (5 x  + 2 x - 8)
> g:=diff(f,x);
                                  2           4
                      g := 5 (5 x  + 2 x - 8)  (10 x + 2)
> int(g,x);
           10          9           8           7           6          5
     3125 x   + 6250 x  - 20000 x  - 38000 x  + 56400 x  + 89632 x

                4          3           2
        -90240 x  - 97280 x  + 81920 x  + 40960 x

> f:=(x-5)/(x^2+1);
                                          x - 5
                                   f := ------
                                          2
                                         x  + 1

> g:=diff(f,x);
```

```
                           1        (x - 5) x
               g := ------ - 2 ---------
                          2          2     2
                         x  + 1     (x  + 1)

> int(g,x);
                                 - 2 x + 10
                         - 1/2 ----------
                                    2
                                   x  + 1

> f:=sin(cos(x))^3;
                                              3
                          f := sin(cos(x))

> g:=diff(f,x);
                                        2
          g := - 3 sin(cos(x))  cos(cos(x)) sin(x)

> int(g,x);
                                           3
                             sin(cos(x))

> f:=3*x^3+4*x+1;
                                  3
                        f := 3 x  + 4 x + 1

> int(f,x);
                                4       2
                         3/4 x  + 2 x  + x

> f:=cos(x)^4*sin(x);
                                      4
                        f := cos(x)  sin(x)

> int(f, x=0 .. Pi);
                                2/5
```

9. The outputs associated with the `plot` commands are given at the end of this section, in the order in which that they are produced.

```
> a:=sin(x)+cos(x)^2;
                                                  2
                      a := sin(x) + cos(x)

> plot(a, x=-2*Pi .. 2*Pi);
> b:=diff(a,x);
> plot({a,b}, x=-2*Pi .. 2*Pi);
```

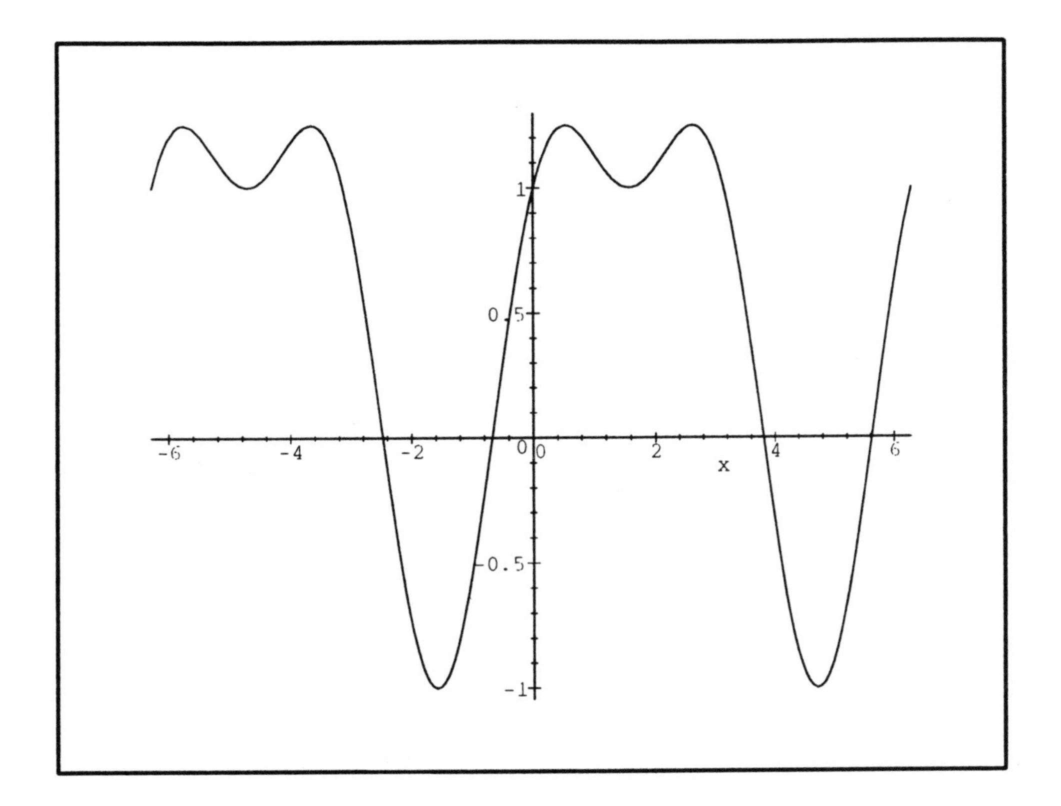
1
0.5
-6
-4
-2
0
0
2
4
6
x
-0.5
-1

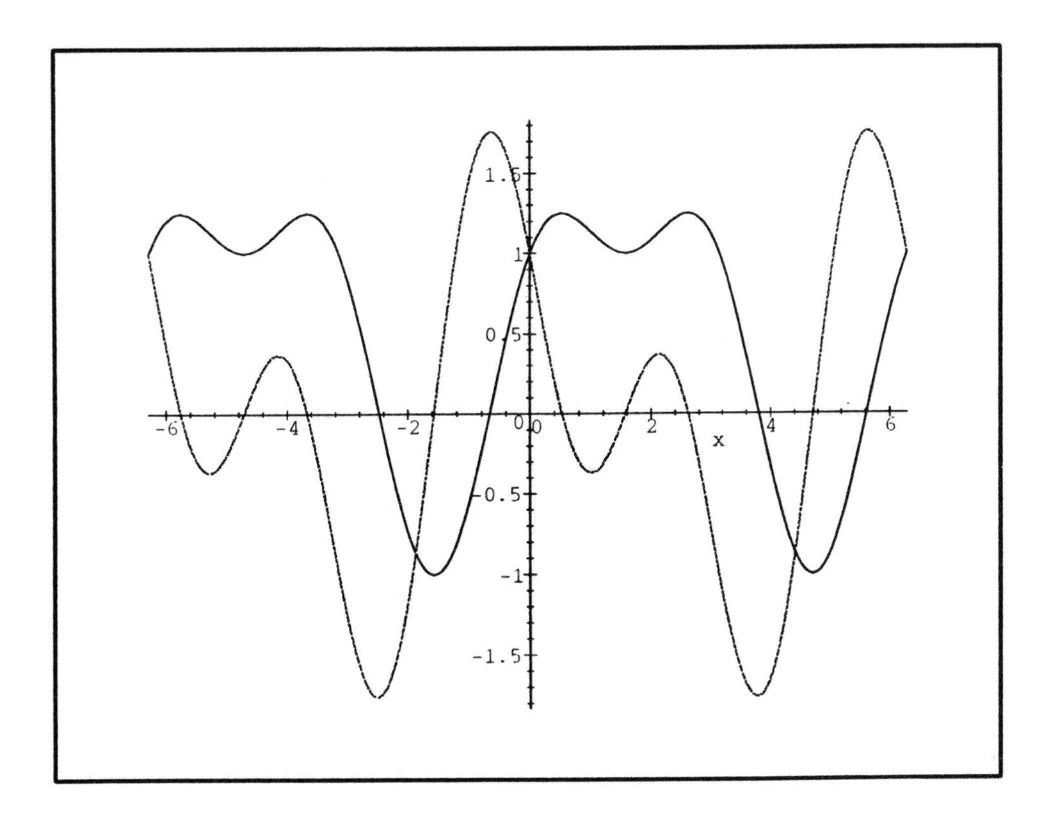
1.5
1
0.5
-6
-4
-2
0
0
2
4
6
x
-0.5
-1
-1.5

## A.6 Exercises

As you obtain answers to the following questions from *Maple*, record your complete interaction (what you type and *Maple*'s response to what you type, with the exceptions of errors).

1. Obtain the first 20 digits of the decimal representation of $\pi$.

2. What is the highest power of 10 that divides 150! ?

3. Enter the polynomial $2x^3 + 9x^2 - 5x - 42$. (You may want to give this expression a name for future reference.)

   (a) Graph the polynomial between $x = -10$ and $x = 10$ and estimate its roots from the graph.

   (b) Graph the polynomial again, this time between $x = -4.5$ and $x = 2.5$. What are your estimates now?

   (c) Find the exact roots of the polynomial. Are they close to your answers to Part (b)?

4. Enter $\sin^2 x + \cos x$ into *Maple*.

   (a) Determine, in decimal form, its value when $x = 1$.

   (b) Estimate the value of $x$ between 2 and 3, accurate to two decimal places, where the graph of $\sin^2 x + \cos x$ crosses the $x$-axis. You may need to "zoom in" on this point in successive graphs (this can be done by reissuing your plot command with smaller interval specifications).

5. Enter the expression $y = x^4 + 6x^3 - 20x^2 - 150x - 100$ into *Maple*.

   (a) Determine the value of $y$ in decimal form, when $x = -\pi$.

   (b) Graph $y$ and its derivative, $y'$, on one set of axes. Explain how you can tell which is the graph of $y$ and which is the graph of $y'$.

6. Enter the expression $f = \dfrac{x}{(x^2+2)^3}$ into *Maple*.

   (a) Find $\int_0^1 f\,dx$, $\int_0^2 f\,dx$, $\int_0^3 f\,dx$,

   (b) Find $g = \int_0^t f\,dx$, as an expression. When $t$ is chosen to be 1, 2, and 3, do these values agree with those obtained in Part (a)?

   (c) Find an antiderivative of $f$. How does this compare with $g$ that you got in Part (b)? Explain.

# Appendix B

# The Statistics Supplement to Maple

The statistics supplement to *Maple* consists of about 130 procedures written specifically to promote explorations of probabilistic and statistical concepts. Here we describe in some detail how each of these procedures is to be used.

Because of the level of detail, we **do not** recommend that this Appendix be read carefully and in its entirety. It would be best if on an initial perusal, the reader obtains a general sense of the types of procedures that are available and how these procedures are used. Later, when a need for a particular procedure arises, more specific information can be obtained either through the on-line help system (`? procedure name`) or by referring to the appropriate portion of this Appendix.

## B.1 The List Data Structure

Before you consider the details of how to use the statistics supplement, it is important that you understand the data structures on which almost all the procedures operate. The "list" structure that is part of the *Maple* system is the underlying data structure that is used in almost all the procedures of the supplement. A list in *Maple* consists of "`[`", followed by a list of objects separated by commas, followed by "`]`". Thus,

```
L := [1,2,3];
```

defines a list (named `L`) consisting of the numbers 1, 2, and 3. The entries of a list need not be numbers; in fact, they do not even have to be of the same type. The statement

```
M := [x,5,Pi,1+5*x*y];
```

gives the name `M` to a list with symbolic as well as numeric entries. It is also possible, and sometimes desirable, to construct a list consisting of lists as in

```
N := [[1,2,3],[2,1,5],[6,5,4,3,2,1]];
```

In this case the first entry of N is the list with entries 1, 2, and 3.

A specific entry of a list can be accessed through indexing; `L[3]`, for example, refers to the third entry of the list `L`. In the case of the List `N`, `N[1]` is itself the list `[1, 2, 3]`. You can access the second entry of `N[1]` either through `X := N[1];` followed by `X[2];` or more directly through `N[1][2];`. Thus, if you want the sum of the elements in the third sublist of `N`, you can give the command

```
sum(N[3][i],i = 1 .. 6);
```

Since you will often use large lists (lists with many entries), you will need to extract some basic information from a list and be able to perform various operations on lists. One of the most commonly needed attributes of a list is the number of entries in it. This is obtained by `nops(L);`, where `L` is a list. For the list `N` of the previous paragraph, `nops(N);` gives 3, whereas `nops(N[3]);` gives 6.

The *Maple* `seq` command can be used to perform a variety of operations on lists. For example, if you have a list, `L`, and want to construct another list, `M`, whose entries consist of the squares of the entries in `L`, you can use

```
M := [seq(L[i]^2,i = 1 .. nops(L))];
```

The *Maple* `seq` command has an alternative form that does not require an index `i`. It is useful when only the numbers inside the list are important. In the form

```
M := [seq(x^2,x = L)];
```

`x` represents an element of the list `L`. The sequence then contains the square of each of the elements of `L`.

## B.2 Procedures for Descriptive Statistics

- `Mean(L)`: This procedure returns the mean (arithmetic average) of its argument. If the argument is a list, then the mean of the entries of the list is returned. For example,

  ```
  M := [1,2,3,4];
  Mean(M);
  ```

  will produce `5/2` and

  ```
  N := [x,3,1+x];
  Mean(N);
  ```

will produce `2/3 x + 4/3`. This procedure will also accept a *frequency list* as its argument. A frequency list is a list of lists of the form `[ [x1, f1], [x2, f2], ..., [xk, fk] ]`, indicating the occurrence of `x1` `f1` times, `x2` `f2` times, etc.

- `Variance(L)`: The argument `L` must either be a list or a frequency list (defined above). This procedure returns the variance of the list `L` or the variance associated with the frequency list. For the lists `M` and `N` defined above, `Variance(M);` and `Variance(N);` will yield, respectively, `5/3` and

```
           2                     2              2
    1/2 x   + 9/2 + 1/2 (1 + x)   - 1/6 (2 x + 4)
```

- `StDev(L)`: The argument `L` must either be a list or a frequency list (defined in `Mean` above). This procedure returns the standard deviation of the list `L` or the standard deviation associated with the frequency list.

- `Max(L)`: Returns the maximum value of the entries of the list `L`. If the list consists of numeric values, the result will be numeric; if the list has symbolic entries then the result will be in symbolic form. For the lists `M` and `N` defined above, `Max(M);` and `Max(N);` will give `4` and

```
    max(3, 1 + x)
```

- `Min(L)`: Returns the minimum value of the entries of the list `L`.

- `Range(L)`: Returns the range (`Max(L) - Min(L)`) of the entries of the list `L`.

- `Percentile(L, p)`: Returns the ($100p$)th percentile of the entries of the list `L`. For example, `Percentile(L,0.50);` will produce the 50th percentile or median and `Percentile(L,0.25);` will produce the first quartile. This procedure requires that the entries of `L` be numeric.

- `Median(L)`: Returns the median of the list `L`. This is a special case of `Percentile`.

- `Skewness(L)`: Returns the skewness (third moment about the mean) of the data in the list `L`. Skewness is used as a measure of "symmetry."

- `Kurtosis(L)`: Returns the kurtosis (fourth moment about the mean) of the data in the list `L`. Kurtosis is used as a measure of "tail weight."

- `Freq(L, xmin .. xmax)`: Produces a list of the frequencies of `xmin,xmin+1,...,xmax` in `L`. The entries of `L` and `xmin, xmax` must be numeric. For example, if `L` is defined by

```
    L := [2,3,3,4,2,7,4,3,2,6];
```

then `F := Freq(L,1 .. 7);` will give

```
F := [0, 3, 3, 2, 0, 1, 1]
```

which indicates that within the list L, the frequencies of 1, 2, 3, 4, 5, 6, and 7 are 0, 3, 3, 2, 0, 1, and 1, respectively.

- `Locate(L, x)`: Returns a list of the locations within the list L that have the entry x. For example, for the list L defined above, `Locate(L,3);` will yield

  ```
  [2, 3, 8]
  ```

  indicating that the second, third, and eighth entries of P are 3's.

- `ClassFreq(X, xmin .. xmax, NumInt)`: Groups the entries of the list X into NumInt equal-sized intervals. In general, xmin should equal the lowest class boundary and xmax should equal the largest class boundary. It should be true that `xmin <= Min(X)` and `xmax >= Max(X)`. If ClassFreq is used with one argument, as in `ClassFreq(X)`, then the actual minimum and maximum values of X will be used and NumInt will be set to 8. If you want observations that are less than xmin to be counted in the first class and observations that are greater than xmax to be counted in the last class, first construct the list Y and then apply ClassFreq to Y where Y can be constructed using

  ```
  Y := [seq(max(min(X[k],xmax),xmin),k = 1 .. n)]:
  ```

  or

  ```
  Y := [seq(max(min(x,xmax),xmin),x = X)]:
  ```

- `RunningSum(L)`: Produces a list of the successive sums of the entries in the list L. The result of `RunningSum(P);`, where P was defined above, will be

  ```
  [2, 5, 8, 12, 14, 21, 25, 28, 30, 36]
  ```

## B.3 Random Sample Generators

- `randomize()`: The procedures described in this section generate random samples from continuous and discrete distributions. These procedures will produce identical results if they are invoked in the same manner and in the same order in successive *Maple* sessions. The `randomize()` procedure modifies the seed of the built-in *Maple* random number generator (the modified seed is actually printed on the screen) in order to produce different random sequences in successive *Maple* sessions.

- `RNG(n)`: Generates a list of `n` random numbers from $[0, 1)$. The result of `RNG(4);` will be similar to

```
[.427419669081, .321110693270, .343633073697, .474256143563]
```

- `Die(m, n)`: Generates a list of `n` random outcomes of the roll of an `m`-sided die. `Die(6,12);` will produce something like

```
[5, 3, 6, 3, 2, 2, 2, 4, 4, 3, 3, 2]
```

- `DiscreteS(L, n)` or `DiscreteS(expr, a .. b, n)`: Generates a list of random observations from a discrete distribution defined through a list or through an expression. In the first case, suppose the random variable $X$ assumes values $c_1, c_2, \ldots, c_m$ with respective probabilities $p_1, p_2, \ldots, p_m$. If `L` is the *probability list* of $X$ (i.e., the list `[c1, p1, c2, p2, ..., cm, pm]`), then `DiscreteS(L, n)` will produce `n` random observations of $X$. For example,

```
L := [1,1/10,2,2/10,3,3/10,4,4/10]:
X := DiscreteS(L,10);
```

  will result in something like

```
X := [4, 3, 4, 4, 4, 2, 2, 3, 3, 2]
```

  If `expr` is a discrete p.d.f. defined on `a,a+1,...,b`, then a random sample of size `n` from this distribution can be obtained through `DiscreteS(expr,a .. b,n);`. So another way of generating a random sample from the distribution described by `L` would be the following:

```
y := x/10;
X := DiscreteS(y,1 .. 4,10);
```

- `ContinuousS(expr, a .. b, n)`: If the expression, `expr`, gives the p.d.f. of a continuous random variable on the interval `[a,b]`, then `n` random observations from this distribution can be generated using `ContinuousS(expr,a .. b,n);`. For example, to generate a random sample of size 5 from the distribution with p.d.f. $f(x) = (3/2)x^2$, $-1 \leq x \leq 1$, you could use

```
f := (3/2)*x^2;
X := ContinuousS(f,-1 .. 1,4);
```

  This would result in something like

```
X := [.8655989594, .9942783194, -.9141777341, -.6816613617]
```

The following procedures generate lists of `n` random observations from specific distributions. In all cases, the parameters of the distribution are given first (from the mnemonics used, you should be able to tell what each parameter stands for) and the last argument is the number of desired observations. You can use the `help` command to get more detailed information.

- `BernoulliS(p, n)`
- `BetaS(alpha, beta, n)`
- `BinomialS(N, p, n)`
- `BivariateNormalS(muX, varX, muY, varY, rho, n)`
- `CauchyS(n)`
- `ChisquareS(r, n)`
- `DiscUniformS(a .. b, n)`
- `ExponentialS(theta, n)`
- `FS(nu1, nu2, n)`
- `GammaS(alpha, theta, n)`
- `GeometricS(p, n)`
- `HypergeometricS(n1, n2, r, n)`
- `LogisticS(n)`
- `NegBinomialS(r, p, n)`
- `NormalS(mu, var, n)`
- `PoissonS(lambda, n)`
- `TS(nu, n)`
- `UniformS(a .. b, n)`

## B.4 Plotting Routines

- `ProbHist(L)` or `ProbHist(expr, xmin .. xmax)`: This procedure displays probability histograms for discrete distributions. The distribution may be defined through an expression representing its p.d.f. or through a list (see `DiscreteS` in the previous section).

- `PlotDiscCDF(expr, xmin .. xmax)`: Produces a graph of the distribution function of the discrete random variable with p.d.f. `expr` (defined in terms of a variable, say `x`) for `x = xmin, xmin+1, ..., xmax`.

- `Histogram(X)` or `Histogram(X, xmin .. xmax, NumInt)`: This procedure plots a histogram with `NumInt` equal-sized intervals for the data defined by the list `X`. It is necessary that `xmin <= Min(X)` and `xmax >= Max(X)`. In general, `xmin` should equal the lowest class boundary and `xmax` should equal the highest class boundary. If `Histogram` is used with one argument, as in `Histogram(X)`, then the actual minimum and maximum values of `X` will be used and `NumInt` will be set to 8. If you want observations that are less than `xmin` to be counted in the first class and observations that are greater than `xmax` to be counted in the last class, then make a histogram of the list `Y` where

  ```
  Y := [seq(max(min(X[k],xmax),xmin),k = 1 .. n)];
  ```

  or

  ```
  Y := [seq(max(min(x,xmax),xmin),x = X)];
  ```

- `PlotEmpPDF(X)` or `PlotEmpPDF(X, xmin .. xmax, NumInt)`: Displays the empirical p.d.f. for the list `X`. It is necessary that `xmin <= Min(X)` and `xmax >= Max(X)`. If `PlotEmpPDF` is used with one argument, as in `PlotEmpPDF(X)`, then the actual minimum and maximum values of `X` will be used and `NumInt` will be set to 8.

- `PlotEmpCDF(X, xmin .. xmax)`: Plots the empirical distribution function of the list of numbers given in `X` over the interval [`xmin, xmax`]. If all of `X` is not in this interval, then the interval will be expanded to cover `X`.

- `Ogive(X, xmin .. xmax, NumInt)`: Displays the ogive for the data described by the list `X`.

- `PlotRunningAverage(L)`: Produces a graph of the running averages in the list `L`.

- `TimePlot(A1, A2, ..., Ak)`: Produces a time series plot of the data represented by the lists `A1, A2, ..., Ak`. **Note: The ... is not part of the syntax of this procedure; it only designates that the procedure can have any number of arguments.**

- `StemLeaf(L)` or `StemLeaf(L, NoDigits)` or `StemLeaf(L, NoDigits, NoRows)`: The list `L` consists of the data for which this procedure will produce a stem-and-leaf display. The second argument `NoDigits`, if present, specifies the number of digits that will be used on the right (i.e., the leaf part); if `NoDigits` is not present, one to three digits will be used, depending on the data. The third argument `NoRows`, if present, specifies the number of rows that will be used in the display; if `NoRows` is not used, the number of rows will be about $\sqrt{n}$ where $n$ is the number of elements in `L`.

- `BoxWhisker(A1, A2, ..., Ak)`: Produces a box-and-whisker plot for each set of data represented by the lists `A1, A2, ..., Ak`. **Note: The ... is not part of the syntax of this procedure; it only designates that the procedure can have any number of arguments.**

- `QQ(A)` or `QQ(X, Y)`: `A` or `X` and `Y` are as described in `ScatPlot`, except that in the case of `X` and `Y` the two lists need not have the same length. This procedure produces a *q-q* (quantile-quantile) plot of the points in `A` or of the lists `X` and `Y`.

- `XbarChart(ListOfSamples)` or `XbarChart(ListOfMeans, SampleSize, MeanOfStdDevs)`: `XbarChart` displays an $\bar{x}$-chart. If only one argument `ListOfSamples` is used, it must be a list of lists, each entry of which is a sample. If three arguments are used, the first must be a list of the means of samples, each sample of size `SampleSize`. The third argument, `MeanOfStDevs` must be the mean of the standard deviations of the samples.

## B.5 Regression

- `Correlation(L)` or `Correlation(X, Y)`: Returns the correlation coefficient of the points $(x_1, y_1)$, $(x_2, y_2)$, ... , $(x_n, y_n)$. If only one argument, `L`, is specified, then it must be a list of lists of the form `[[x1,y1], [x2,y2], ...,[xn,yn]]`. If two arguments, `X` and `Y` are used, then `X` must be defined as the list of $x$-coordinates, `[x1, x2, ..., xn]` and `Y` must be defined as the list of corresponding $y$-coordinates, `[y1, y2, ..., yn]`. For example,

```
X := [1,2,3,4,5];
Y := [2,3,1,5,4];
r := Correlation(X,Y);
```

  or

```
XY := [[1,2],[2,3],[3,1],[4,5],[5,4]];
r := Correlation(XY);
```

will give

```
r := .6000000000
```

- `LinReg(L, x)` or `LinReg(X, Y, x)`: `L` or alternatively `X` and `Y` must be of the type described in `Correlation` above and `x` must be a variable name. `LinReg(L,x);` or `LinReg(X,Y,x)` returns the `Y` on `X` linear regression equation in the variable `x`. For `X, Y,` and `XY`, defined above,

  ```
  y := LinReg(X,Y,x);
  ```

  or

  ```
  y := LinReg(XY,x);
  ```

  will produce

  ```
  y := 1.200000000 + .6000000000 x
  ```

- `PolyReg(L, deg, x)` or `PolyReg(X, Y, deg, x)`: `L, X, Y` and `x` are as described in `LinReg` and `deg` is the degree of a polynomial. This procedure finds a least squares polynomial of degree `deg` for the points given in `X` and `Y` (or in `L`). For `X, Y,` and `XY` defined above,

  ```
  y := PolyReg(X,Y,3,x);
  ```

  or

  ```
  y := PolyReg(XY,3,x);
  ```

  will produce

  ```
                88         23   2          3
  y := 5 - ---- x + ---- x  - 1/6 x
                21         14
  ```

- `ScatPlot(A)` or `ScatPlot(X, Y)`: Produces a scatter plot of the $n$ points

  $$(x_1, y_1),\ (x_2, y_2),\ \ldots,\ (x_n, y_n).$$

  If only one argument `A` is specified, then `A` must be a list of lists of the form

  ```
  A := [[x1,y1],[x2,y2],...,[xn,yn]];
  ```

If two arguments X and Y are used, then X and Y must be lists of the respective $x$-coordinates and $y$-coordinates:

```
X := [x1,x2,...,xn];
Y := [y1,y2,...,yn];
```

- ScatPlotLine(A) or ScatPlotLine(X, Y): For a list of lists A or lists X and Y of the type described in ScatPlot above, this procedure produces the same scatter plot as ScatPlot. In this case the regression line of the points is superimposed on the scatter plot.

- PlotPolyReg(A, deg) or PlotPolyReg(X, Y, deg): For a list of lists A or lists X and Y of the type described in ScatPlot above, this procedure produces the same scatter plot as ScatPlot. In this case the polynomial regression line of the points is superimposed on the scatter plot. The degree of the polynomial is deg.

- Residuals(A, deg) or Residuals(X, Y, deg): For a list of lists A or lists X and Y of the type described in ScatPlot above, this procedure graphically displays the residuals associated with a polynomial fit of degree deg and then gives the residual values in tabular form.

- RegBand(A, ConfLev, Type) or RegBand(X, Y, ConfLev, Type): For a list of lists A or lists X and Y of the type described in ScatPlot above, this procedure will produce the same scatter plot and regression line as ScatPlotLine. In this case, depending on whether Type is C or P, a ConfLev% confidence band or prediction band will be included.

## B.6 p.d.f.s of Some Distributions

The p.d.f.s of a number of commonly used distributions are made available by the procedures of this section. The names of these procedures consist of the distribution name (as always with capitalized first letter) followed by PDF. The arguments consist of the distribution parameters (from the mnemonics used, you should be able to tell what each parameter stands for). The last argument is a variable name; the p.d.f. will be expressed in terms of this variable. **Note: The parameters of the distribution can be in numeric or symbolic form.**

- BernoulliPDF(p, x): If p is numerical as in BernoulliPDF(0.2,t);, the result will be

```
  t   (1 - t)
.2  .8
```

If `f := BernoulliPDF(p,t);` is used,

```
            t          (1 - t)
    f := p   (1 - p)
```

will be returned.

- `BetaPDF(alpha, beta, x)`
- `BinomialPDF(N, p, x)`
- `BivariateNormalPDF(muX, varX, x, muY, varY, y, rho)`: Since this is a bivariate distribution, two variable names `x` and `y` are specified.
- `CauchyPDF(x)`
- `ChisquarePDF(r, x)`
- `DoubleExponentialPDF(theta, sigma, x)`
- `ExponentialPDF(theta, x)`
- `FPDF(nu1, nu2, x)`
- `GammaPDF(alpha, theta, x)`
- `GeometricPDF(p, x)`
- `HypergeometricPDF(n1, n2, r, x)`
- `LogisticPDF(x)`
- `NegBinomialPDF(r, p, x)`
- `NormalPDF(mu, var, x)`
- `PoissonPDF(lambda, x)`
- `TPDF(nu, x)`
- `UniformPDF(a .. b, x)`
- `WeibullPDF(alpha, beta, theta, x)`

## B.7 c.d.f.s of Some Distributions

The c.d.f.s of a number of commonly used distributions are made available by the procedures of this section. The names of these procedures consist of the distribution name (as always with capitalized first letter) followed by `CDF`. The arguments consist of the distribution parameters (from the mnemonics used, you should be able to tell what each parameter stands for). The last argument is a variable name; the c.d.f. will be expressed in terms of this variable. **Note: The parameters of the distribution can be in numeric or symbolic form.**

- `BernoulliCDF(p, x)`: Returns the c.d.f. of the Bernoulli distribution. If used with a numeric value of `p`, as in `BernoulliCDF(0.2, t);` the following is returned.

```
                                (t + 1.)
  - 1.066666667 .2500000000              + 1.066666667
```

  The result of `F := BernoulliCDF(p, t);` will be

```
                      2 /            p     \(t + 1.)
           ( - 1. + p)  | - 1. ---------|                      2
                        \        - 1. + p/               ( - 1. + p)
  F := ---------------------------------------- - 1. ------------
                        2. p - 1.                       2. p - 1.
```

  This is correct; however, it is far more complicated than it needs to be. We expect that future releases of *Maple* will produce a simpler answer.

- `BetaCDF(alpha, beta, x)`
- `BinomialCDF(N, p, x)`
- `ChisquareCDF(r, x)`
- `ExponentialCDF(theta, x)`
- `FCDF(nu1, nu2, x)`
- `GammaCDF(alpha, theta, x)`
- `GeometricCDF(p, x)`
- `NegBinomialCDF(r, p, x)`
- `NormalCDF(mu, var, x)`
- `PoissonCDF(lambda, x)`
- `TCDF(nu, x)`
- `UniformCDF(a .. b, x)`

## B.8 Percentiles of Some Distributions

The percentiles of a number of commonly used distributions are made available by the procedures of this section. The names of these procedures consist of the distribution name (as always with capitalized first letter) followed by `P`. The arguments consist of the distribution parameters (from the mnemonics used, you should be able to tell what each parameter stands for). The last argument is the percentile value (if it is numeric, it must be between 0 and 1). The parameters of the distribution, as well as the percentile, can be in numeric or symbolic form. If any of the arguments is in symbolic form, the result is left unevaluated. **Note: Since the evaluation of percentiles involves the solution of integral equations that generally cannot be expressed in closed form, the percentile procedures do not always provide an answer.**

- `ChisquareP(r, p)`: This procedure returns percentiles of chi-square distributions with `r` degrees of freedom. `ChisquareP(5,0.9);` will produce `9.236356900`. If `ChisquareP(df,0.9);` is used, then `ChisquareP(df, .9)` is returned. Similarly, if `ChisquareP(5,p);` is used, then `ChisquareP(5, p)` is returned.
- `BetaP(alpha, beta, p)`
- `ExponentialP(theta, p)`
- `FP(nu1, nu2, p)`
- `GammaP(alpha, theta, p)`
- `NormalP(mu, var, p)`
- `TP(nu, p)`
- `UniformP(a .. b, p)`

## B.9 Samples from Sampling Distributions

The procedures described in Section B.3, together with some computation, can be used to produce random samples from sampling distributions. For example, the following sequence can be used to obtain a list of 100 means of samples of size 5 from the uniform distribution on the interval [2, 4].

```
samples := [seq(UniformS(2 .. 4,5),i = 1 .. 100)];
means := [seq(Mean(samples[i]),i = 1 .. 100)];
```

The procedures of this section make this process more convenient, and in a number of cases, computationally more efficient.

- `UniformMeanS(a .. b, n, m)`: This procedure generates `m` means of random samples of size `n` from $U(a, b)$. The result of `UniformMeanS(2 .. 4,10,4);` will be something like

```
[2.813766232, 3.146085917, 3.212233169, 2.952763787]
```

- `UniformMedianS(a .. b, n, m)`: Generates `m` medians of samples of size `n` from uniform distributions.

- `UniformSumS(a .. b, n, m)`: Generates `m` sums of samples of size `n` from uniform distributions.

- `NormalMeanS(mu, var, n, m)`: Generates `m` means of samples of size `n` from normal distributions.

- `NormalMedianS(mu, var, n, m)`: Generates `m` medians of samples of size `n` from normal distributions.

- `NormalVarianceS(mu, var, n, m)`: Generates `m` variances of samples of size `n` from normal distributions.

- `NormalSumS(mu, var, n, m)`: Generates `m` sums of samples of size `n` from normal distributions.

- `NormTransVarS(mu, var, n, m)`: Generates `m` observations of $(n-1)S^2/\sigma^2$ for random samples of size `n` from normal distributions.

- `ExponentialSumS(theta, n, m)`: Generates `m` sums of samples of size `n` from exponential distributions.

## B.10 Confidence Intervals

- `ConfIntMean(ListOfSamples, ConfLev)` or `ConfIntMean(ListOfSamples, ConfLev, variance)`: `ListOfSamples` is a list of lists, each entry of which is a sample. All the samples in `ListOfSamples` must be of the same size. This procedure produces `ConfLev%` confidence intervals for the mean for each of the samples in `ListOfSamples`. The third argument, `variance`, is optional. If it is present, it will be assumed that the population variance is known and it is equal to `variance`. If only two arguments are used, it will be assumed that the population variance is unknown.

- `ConfIntVar(ListOfSamples, ConfLev)`: Produces `ConfLev%` confidence intervals for the variance for each of the samples in `ListOfSamples` (the structure of `ListOfSamples` is as described above).

- `ConfIntProp(ListOfSamples, ConfLev)`: Produces `ConfLev`% confidence intervals for $p$, the probability of success in a Bernoulli trial (`ListOfSamples` is as described above and in this case the samples must consist of 0's and 1's).

- `ConfIntSuc(ListOfIntervals, v)`: The first argument, `ListOfIntervals`, is a list of lists of the form `[ [a1, b1], [a2, b2], ..., [ak, bk] ]` where each entry (e.g., `[ai, bi]`) is a confidence interval (probably produced by `ConfIntMean` or `ConfIntVar` or `ConfIntProp`). This procedure returns the number of "successes," i.e., the number of intervals that contain the parameter `v`.

- `ConfIntAvLen(ListOfIntervals)`: `ListOfIntervals`, the only argument, is as described above in `ConfIntSuc`. The procedure returns the average length of the intervals in `ListOfIntervals`.

- `ConfIntPlot(ListOfIntervals)` or `ConfIntPlot(ListOfIntervals, v)`: The argument, `ListOfIntervals`, is as described in `ConfIntSuc`. This procedure produces a plot of the intervals in `ListOfIntervals`. The second argument is optional; if used, a vertical line at the parameter `v` will be superimposed on the plot of the confidence intervals.

# B.11 Analysis of Variance

- `Anova1(L)`: This procedure performs a one-factor analysis of variance. The argument `L` is a list of lists, where each sublist consists of the measurements associated with one of the classifications (factor levels). For example, to use `Anova1` with measurements 13, 8, 9; 15, 11, 13; 11, 15, 10, representing three observations in each of three classifications, you would define `L` and then invoke `Anova1` (the *Maple* commands and the output from `Anova1` are given below).

```
L := [[13,8,9],[15,11,13],[8,12,7],[11,15,10]];
Anova1(L);
```

```
              Sum of
              Squares    Degrees of   Mean Sq
Source         (SS)        Freedom      (MS)        F-Ratio    p-value
-----------------------------------------------------------------------
Treat.       30.0000          3       10.0000       1.6000      0.2642
Error        50.0000          8        6.2500
Total        80.0000         11
-----------------------------------------------------------------------
```

- `Anova2s(L)`: This procedure performs a two-factor analysis of variance with a single observation per cell. The argument `L` is a list of lists, where each sublist consists of the measurements associated with the row factor. For example, to use **Anova2s** with row measurements 16, 18, 21, 21; 14, 15, 18, 17; 15, 15, 18, 16, we would define `L` and then invoke **Anova2s** (the *Maple* commands and the output from **Anova2s** are given below).

```
M := [[16,18,21,21],[14,15,18,17],[15,15,18,16]];
Anova2s(M);
```

| Source | Sum of Squares (SS) | Degrees of Freedom | Mean Sq (MS) | F-Ratio | p-value |
|---|---|---|---|---|---|
| Row(A) | 24.0000 | 2 | 12.0000 | 18.0000 | 0.0029 |
| Col(B) | 30.0000 | 3 | 10.0000 | 15.0000 | 0.0034 |
| Error | 4.0000 | 6 | 0.6667 | | |
| Total | 58.0000 | 11 | | | |

- `Anova2m(L)`: This procedure performs a two-factor analysis of variance with multiple observations per cell. The argument `L` is a list of lists of lists, of the form

  ```
  L := [[L11,L12,...,L1c],[L21,L22,...,L2c],...,[Lr1,Lr2,...,Lrc]]:
  ```

  where `Lij` is the list of measurements in row `i` and column `j`. The following illustrates a use of **Anova2m**.

  ```
  N11 := [21,14,16,18]: N12 := [16,20,15,21]: N13 := [15,19,14,14]:
  N14 := [17,17,18,20]: N21 := [20,19,20,18]: N22 := [15,16,17,19]:
  N23 := [16,13,18,16]: N24 := [18,17,21,19]: N31 := [14,13,16,17]:
  N32 := [17,18,16,16]: N33 := [15,16,17,18]: N34 := [16,16,17,18]:
  N:=[[N11,N12,N13,N14],[N21,N22,N23,N24],[N31,N32,N33,N34]]:
  Anova2m(N);
  ```

| Source | Sum of Squares (SS) | Degrees of Freedom | Mean Sq (MS) | F-Ratio | p-value |
|---|---|---|---|---|---|
| Row(A) | 15.7917 | 2 | 7.8958 | 2.1992 | 0.1256 |
| Col(B) | 23.0625 | 3 | 7.6875 | 2.1412 | 0.1121 |
| Int(AB) | 34.8750 | 6 | 5.8125 | 1.6190 | 0.1703 |
| Error | 129.2500 | 36 | 3.5903 | | |
| Total | 202.9792 | 47 | | | |

## B.12 Goodness of Fit Tests

- `ChisquareFit(DataList, CDFexpr, Classes)` or `ChisquareFit(DataList, CDFexpr, Classes, Discrete)`: This procedure performs a chi-square goodness of fit test for the data in the list `DataList` and the distribution whose c.d.f. is given by `CDFexpr`. `Classes` is a list of the form `[a1, a2, ..., ak]` describing classes

$$[a_1, a_2), [a_2, a_3), [a_3, a_4), \ldots, [a_{k-1}, a_k).$$

If `DataList` contains numbers less than `a1` or greater than or equal to `ak` then these are absorbed into the first and last classes, respectively. The fourth argument is optional and if present, it can be anything. Its presence simply indicates that `CDFexpr` is the c.d.f. of a discrete distribution. `ChisquareFit` produces a histogram of the data on the intervals designated by `Classes` with a superimposed probability histogram (in the discrete case) or p.d.f. (in the continuous case). Additionally, `ChisquareFit` gives the chi-square goodness of fit statistic, the degrees of freedom associated with the test, and the $p$-value of the test. The following examples illustrates the use of `ChisquareFit` in the discrete case. Note that if the possible outcomes are the integers `a1, a2, ..., ak`, then the last argument in the list `Classes` should be `ak + 1`.

```
F := x/10;
Dat := Die(10,100);
C := [1, 2, 3, 4, 5, 6, 7, 8, 9, 10, 11];
ChisquareFit(Dat,F,C,Discrete);

F := PoissonCDF(6,x);
Dat := PoissonS(6,300);
C := [2, 3, 4, 5, 6, 7, 8, 9, 10, 11, 12, 13)];
ChisquareFit(Dat,F,C,Discrete);
```

The following example illustrates `ChisquareFit` in the continuous case:

```
F := ExponentialCDF(10,x);
Dat := ExponentialS(10,200);
C := [0, 10, 15, 20, 25, 30, 35, 40];
ChisquareFit(Dat,F,C);
```

- `KSFit(DataList, CDFexpr, xmin .. xmax)`: This procedure does a Kolmogorov-Smirnov goodness of fit test for the data in the list `DataList` and the distribution whose c.d.f. is given by `CDFexpr`. `KSFit` produces, on the interval designated by `xmin .. xmax`, an empirical c.d.f. of the data superimposed with the theoretical c.d.f. given by `CDFexpr`. Additionally, `KSFit` gives the Kolmogorov-Smirnov goodness of fit statistic, and the location where this maximal value is attained. The following illustrates the use of `KSFit`.

```
F := NormalCDF(100,225,x);
Dat := NormalS(100,225,30);
KSFit(Dat,F,50 .. 150);
```

## B.13 Nonparametric Tests

- `SignTest(Bexpr, x=X, y=Y, ...)`: This procedure counts the number of times the items of one or more lists fulfill certain conditions. `Bexpr` is a boolean expression in at least one (unassigned) variable. The remaining arguments are of the form `name=list` (one is needed for each variable in `Bexpr`). For example,

  ```
  SignTest(x > 0.5,x = RNG(50));
  ```

  will determine the number of the 50 random numbers that are greater than 0.5, and

  ```
  SignTest(x > y,x = RNG(50),y = RNG(50));
  ```

  will determine the number of times $x_i > y_i, i = 1, 2, \ldots, 50$.

- `Wilcoxon(L, m)`: Returns the Wilcoxon statistic for testing the median of a sample (`m` is subtracted from each element of the list `L` before ranking). Zero ranks (`L[i]-m = 0`) are deleted and ties are assigned averaged ranks. The following is a typical use of `Wilcoxon(L,m)`:

  ```
  Wilcoxon(RNG(10),0.5);
  ```

- `MedianTest(X, Y, k)`: Counts the number of observations of the list `X` in the lower `k` elements of the list that is obtained by combining and then sorting the lists `X` and `Y`. If `nops(X)+nops(Y)` is even, the third argument `k` may be omitted; in this case, `(nops(X)+nops(Y))/2` will be used in the place of the third argument. Ties are **not** detected.

- `Wilcoxon2(X, Y)`: This procedure performs the two-sample test for equality of medians. It returns the sum of the ranks of the elements of the list `Y` (note difference from above). Ties are given averaged ranks.

- `RunTest(X, Y)` or `RunTest(X, m)`: `RunTest(X, Y)` counts the runs in the combined list of `X` and `Y`. `RunTest(X, m)` tests for randomness and counts the number of runs of observations that are above and below `m`.

## B.14 Miscellaneous Items

- The statistics supplement has a list structure called `Cards` that consists of

  ```
  L := [C1,C2,...,C13,D1,D2,...,D13,H1,H2,...,H13,S1,S2,...,S13];
  ```

  This makes it easy to simulate poker or bridge hands.

- `MarginalRelFreq(A)`: If `A` is a two dimensional list (or equivalently a list of lists) then `MarginalRelFreq` produces `[list1,list2]` where `list1` is a list of column relative frequencies in `A` and `list2` is a list of row relative frequencies in `A`.

- `Craps()`: This procedure simulates a single craps game and returns a list. The first entry of the list is 0 or 1, depending on whether the player loses or wins the game; the remaining entries give the sequence of die rolls that produced the game. For example, a return of `[1, 6, 4, 11, 3, 8, 6]` indicates a win on successive rolls of 6, 4, 11, 3, 8, and 6.

- `FastCraps(n)`: This is a compact version of `Craps();`. It returns the win/loss results of `n` simulated craps games (as 0's and 1's) without the sequences of dice rolls.

- `Contingency(L)`: The argument `L` must be a list of lists, each entry of which is a list of observed frequencies from a contingency table. This procedure prints a table of the expected frequencies and then prints the chi-square statistic and its $p$-value.

- `RandWalk(pn, ps, pe, steps, n)`: Starting at $(0,0)$, an object moves (a fixed unit distance) North, South, East and West directions with probabilities `pn, ps, pe, 1-pn-ps-pe`, respectively. `RandWalk(pn, ps, pe, steps, n)` gives the `n` coordinates of the location of the object after `steps` moves.

- `GraphRandWalk(pn, ps, pe, steps)`: Plots a single path of an object moving in the manner described above.

- `RandWalk2(pn, pe, steps, n)`: Starting at $(0,0)$, an object moves (a fixed unit distance) North-South **and** East-West with probabilities `pn, 1-pn, pe, 1-pe`, respectively. `RandWalk2(pn, pe, steps, n)` gives the `n` coordinates of the location of the object after `steps` moves.

- `GraphRandWalk2(pn, pe, steps)`: Plots a single path of an object moving in the manner described in `RandWalk2`.

- `Convolution(X1, X2)`: The lists `X1` and `X2` are probability lists that represent discrete distributions (to represent the distribution of the roll of a die as a probability list, we define the list as `X1 := [1,1/6,2,1/6,3,1/6,4,1/6,5,1/6,6,1/6];`).

This procedure returns a list that represents the convolution of the distributions represented by `X1` and `X2`. For example, the distribution of the sum of two rolls of a die can be found as follows:

```
X1 := [1,1/6,2,1/6,3,1/6,4,1/6,5,1/6,6,1/6];
X2 := [1,1/6,2,1/6,3,1/6,4,1/6,5,1/6,6,1/6];
Y := Convolution(X1,X2);
```

# Appendix C

# Table of Random Numbers[1]

| | | | | | | | | | |
|---|---|---|---|---|---|---|---|---|---|
| 42742 | 32111 | 34363 | 47426 | 55846 | 74675 | 03206 | 72297 | 60431 | 74558 |
| 25981 | 31008 | 79718 | 03917 | 08843 | 96050 | 81292 | 45375 | 64403 | 92062 |
| 95105 | 14649 | 15559 | 42939 | 52543 | 27260 | 21976 | 67598 | 84547 | 67647 |
| 28134 | 79250 | 75121 | 62836 | 31375 | 00586 | 07481 | 64384 | 13191 | 67208 |
| 13554 | 99164 | 45261 | 74287 | 50254 | 19905 | 82524 | 99190 | 11688 | 34081 |
| 79539 | 62948 | 73645 | 40159 | 17831 | 91483 | 28131 | 45410 | 76903 | 08070 |
| 60513 | 55292 | 89152 | 62032 | 91285 | 52572 | 44747 | 98149 | 41345 | 99299 |
| 15763 | 88250 | 33906 | 06591 | 11952 | 57416 | 15065 | 55341 | 08282 | 35963 |
| 68159 | 70974 | 23110 | 69380 | 18830 | 01082 | 05431 | 54214 | 10621 | 68722 |
| 02581 | 12102 | 66529 | 09740 | 78042 | 98779 | 67420 | 13405 | 75487 | 14081 |
| 34788 | 43360 | 89872 | 48553 | 25505 | 95292 | 64207 | 15491 | 85607 | 68141 |
| 96292 | 87417 | 90595 | 54955 | 08413 | 06706 | 62176 | 22358 | 27357 | 41042 |
| 65950 | 88797 | 23445 | 60639 | 07046 | 32206 | 61232 | 69815 | 78855 | 56805 |
| 98548 | 81056 | 65120 | 22361 | 05971 | 88581 | 20874 | 86349 | 13491 | 89722 |
| 82539 | 90817 | 85250 | 42524 | 94764 | 35125 | 87165 | 02520 | 19464 | 54853 |
| 55900 | 33129 | 25684 | 67779 | 32635 | 73716 | 52141 | 83580 | 37291 | 32444 |
| 53272 | 33313 | 77722 | 29723 | 03514 | 03239 | 04126 | 46484 | 66274 | 05272 |
| 36425 | 82680 | 47811 | 22217 | 16870 | 62071 | 85283 | 63116 | 60280 | 56427 |
| 24911 | 05476 | 32929 | 78412 | 42257 | 74543 | 69352 | 63255 | 08708 | 99629 |
| 18726 | 46891 | 09330 | 83264 | 87673 | 26058 | 06579 | 31694 | 94010 | 58389 |
| 34234 | 77312 | 63864 | 90503 | 93400 | 87173 | 10217 | 29482 | 27344 | 78560 |
| 83372 | 68192 | 01832 | 28985 | 36652 | 19766 | 26787 | 15974 | 99699 | 87689 |
| 39350 | 32749 | 85647 | 89801 | 36917 | 68350 | 52561 | 42237 | 96883 | 99421 |
| 86634 | 93836 | 80227 | 57131 | 76801 | 31217 | 35010 | 07780 | 83620 | 62411 |
| 03562 | 06506 | 31466 | 05690 | 19580 | 59543 | 80707 | 06996 | 84739 | 40708 |
| 94944 | 41600 | 72786 | 08767 | 26929 | 97555 | 70793 | 10996 | 75925 | 73412 |
| 51572 | 60171 | 82933 | 91980 | 78681 | 35320 | 38439 | 07261 | 17191 | 58079 |
| 02808 | 84511 | 51425 | 01467 | 20570 | 85220 | 62758 | 28283 | 03740 | 14268 |
| 24863 | 31880 | 56731 | 32345 | 30348 | 50598 | 34743 | 23241 | 57108 | 31171 |
| 74050 | 35233 | 22960 | 60897 | 73738 | 67017 | 32508 | 77275 | 98104 | 31232 |

[1] The numbers in this table were produced with the *Maple* random number generator, `rand()`. Entries from this table can be used as random digits; also, each group of 5 digits, when multiplied by $10^{-5}$, can be interpreted as a random number from the interval $[0, 1)$.

| | | | | | | | | | |
|---|---|---|---|---|---|---|---|---|---|
| 37004 | 26229 | 76491 | 47471 | 06523 | 80140 | 11644 | 41434 | 43261 | 46271 |
| 14029 | 46116 | 12110 | 92155 | 50997 | 83020 | 62896 | 57903 | 15178 | 16699 |
| 26183 | 96188 | 42083 | 59176 | 27178 | 53660 | 44860 | 17212 | 79645 | 05954 |
| 32791 | 15286 | 12094 | 89940 | 27646 | 11946 | 62305 | 68949 | 81422 | 37144 |
| 47489 | 35453 | 85456 | 65040 | 15086 | 11632 | 92135 | 80264 | 57857 | 17146 |
| 00411 | 69261 | 19017 | 62309 | 93098 | 31076 | 11261 | 13912 | 89675 | 22236 |
| 43336 | 69052 | 32698 | 45356 | 06048 | 00296 | 78694 | 90140 | 39330 | 16184 |
| 29278 | 85819 | 42540 | 74056 | 46107 | 99476 | 05322 | 29067 | 55218 | 60017 |
| 20187 | 03445 | 06118 | 26022 | 63529 | 73963 | 32057 | 70670 | 61298 | 93550 |
| 47592 | 84132 | 18093 | 35037 | 77122 | 07642 | 59322 | 59574 | 80552 | 83645 |
| 90786 | 49505 | 32929 | 42661 | 35853 | 34068 | 83759 | 90415 | 82128 | 99017 |
| 39119 | 97913 | 84827 | 32055 | 14756 | 54223 | 44839 | 67232 | 07853 | 05350 |
| 06496 | 69710 | 99335 | 54265 | 04343 | 46410 | 26866 | 85940 | 51158 | 85069 |
| 79545 | 05823 | 47508 | 31854 | 14464 | 22896 | 34564 | 64415 | 77237 | 32934 |
| 75058 | 35521 | 00665 | 24351 | 43055 | 02460 | 49630 | 63775 | 26070 | 59083 |
| 57273 | 51236 | 14894 | 97257 | 20217 | 59330 | 21864 | 30814 | 49943 | 21478 |
| 24250 | 93922 | 31571 | 22700 | 95659 | 56152 | 64523 | 91200 | 67382 | 49679 |
| 52181 | 11094 | 44508 | 01002 | 75280 | 23645 | 98969 | 13042 | 94379 | 54545 |
| 51445 | 80278 | 48327 | 51794 | 98395 | 17092 | 67405 | 66646 | 01090 | 69629 |
| 63472 | 53930 | 40699 | 44975 | 12423 | 48603 | 81069 | 81034 | 03639 | 65338 |
| 27054 | 98444 | 05663 | 66627 | 27244 | 97581 | 57995 | 02845 | 15946 | 35118 |
| 22301 | 37919 | 02491 | 71116 | 18810 | 46363 | 54344 | 53676 | 22887 | 71937 |
| 66302 | 68733 | 79318 | 24242 | 64741 | 03763 | 64527 | 70704 | 64872 | 67792 |
| 46531 | 42328 | 83754 | 98895 | 23573 | 36986 | 06711 | 65348 | 61427 | 05603 |
| 04925 | 42150 | 68211 | 29889 | 19676 | 38102 | 47771 | 81506 | 69163 | 23594 |
| 65570 | 67931 | 65653 | 74165 | 47022 | 19275 | 50299 | 02372 | 93916 | 08150 |
| 65814 | 73063 | 32187 | 83812 | 01699 | 84976 | 64154 | 19520 | 71690 | 82050 |
| 64266 | 47446 | 49499 | 99970 | 15000 | 07457 | 61908 | 61481 | 64747 | 05263 |
| 64642 | 93630 | 28915 | 91827 | 79097 | 45057 | 13448 | 84889 | 02290 | 33317 |
| 29363 | 92567 | 63158 | 06741 | 24195 | 53472 | 69651 | 09584 | 03990 | 94458 |
| 83373 | 43687 | 78825 | 40480 | 20007 | 28658 | 56565 | 05350 | 76955 | 33172 |
| 58052 | 05189 | 40165 | 69672 | 58013 | 18992 | 26362 | 40406 | 42771 | 11262 |
| 44095 | 84162 | 19329 | 98991 | 26305 | 96638 | 69397 | 40279 | 00910 | 46687 |
| 86000 | 66365 | 30095 | 95190 | 03966 | 71128 | 92246 | 77992 | 17546 | 84957 |
| 51368 | 38098 | 93367 | 26117 | 95145 | 19690 | 88826 | 60171 | 65904 | 40268 |
| 73280 | 60199 | 44518 | 44430 | 30861 | 86198 | 44146 | 53836 | 73879 | 34523 |
| 77887 | 68197 | 50755 | 32019 | 35782 | 95583 | 57979 | 76052 | 74099 | 77812 |
| 36945 | 64947 | 94139 | 48407 | 17883 | 79589 | 49949 | 47637 | 58731 | 89870 |
| 49297 | 15217 | 61941 | 70656 | 84850 | 71815 | 75531 | 31647 | 91748 | 10978 |
| 33605 | 00242 | 66451 | 25313 | 96949 | 75640 | 18320 | 50855 | 10800 | 58209 |
| 26413 | 46955 | 48369 | 04762 | 50380 | 46705 | 53169 | 68153 | 49026 | 67725 |
| 78850 | 36368 | 43540 | 74774 | 93151 | 74186 | 32389 | 44371 | 26984 | 28409 |
| 20609 | 65382 | 24815 | 52204 | 71784 | 87703 | 50227 | 09130 | 07815 | 35106 |
| 86177 | 74760 | 62400 | 58629 | 57896 | 44997 | 32792 | 67813 | 75896 | 86446 |
| 54594 | 70998 | 52715 | 53074 | 49160 | 63019 | 19063 | 42098 | 30873 | 55100 |
| 01667 | 29540 | 50427 | 06815 | 29483 | 84311 | 74131 | 55755 | 77396 | 20628 |
| 94840 | 50598 | 13810 | 27834 | 91930 | 99889 | 20577 | 06981 | 88309 | 62074 |
| 92650 | 92115 | 48273 | 42874 | 96876 | 96450 | 32900 | 87629 | 94431 | 11747 |
| 42686 | 08973 | 09315 | 01450 | 49664 | 96094 | 49121 | 98721 | 26144 | 42003 |
| 94925 | 07692 | 20936 | 25507 | 64793 | 31175 | 04267 | 71279 | 91961 | 47806 |